EVALUATION METHODS FOR BUSINESS FORM COPIES

A symposium
presented at the
Seventy-third Annual Meeting
AMERICAN SOCIETY FOR
TESTING AND MATERIALS
Toronto, Ont., Canada, 21-26 June 1970

ASTM SPECIAL TECHNICAL PUBLICATION 482

List price $4.00

AMERICAN SOCIETY FOR TESTING AND MATERIALS
1916 Race Street, Philadelphia, Pa. 19103

Library of Congress Catalog Card Number: 71-137452
ISBN 0-8031-0067-1

NOTE

The Society is not responsible, as a body, for the statements and opinions advanced in this publication.

Printed in Baltimore, Md.
January 1971

FOREWORD

The Symposium on Evaluation Methods for Business Form Copies was given at the Seventy-third Annual Meeting of the American Society for Testing and Materials held in Toronto, Ont., Canada, 21–26 June 1970. The sponsor of this symposium was Committee F-5 on Carbon Paper, Inked Ribbon, and Similar Image Producing Products, D. M. Davidson, Mazer Corp., formerly Standard Register Company, presided as symposium chairman.

Related ASTM Publications

Paper and Paperboard, Characteristics, Nomenclature, and Significance of Tests, STP 60-B (1963), $4.00

Contents

Introduction

The quality evaluation of multiform copies falls into the same category as color, odor, and taste measurements since the senses and psychological influences of the observer are involved. Nevertheless, manufacturers and users of business forms have felt the need for instrumental methods and quantitative values in measuring the properties of copies.

In this series of papers, several instruments and measuring concepts are presented. Many of the problems related to this type of product evaluation are discussed, particularly the need for precise specimen preparation and careful reproducibility of results. While only one paper relates to electrostatic copying, the kinship between the older impressioned copy and its younger cousin is unmistakable. A final paper offers a description of the mechanics of ink transfer from carbon paper or ribbon during impact. The theories proposed in this paper represent a significant advance in current concepts and should prove to be very useful to the technology of business forms.

Considerable effort has gone into the development of business copy products, and, while their influence on our lives frequently is not appreciated simply because of their ubiquity, we literally could not live our modern lives without them. Every bill we receive, every item we buy, sell, or ship has a document related to it (prepared, at least, in triplicate) and is a product of the business forms industry.

It is not unusual that this industry, like all others today, is becoming more technically oriented. The materials and processes of production are equally as sophisticated as any other industry. It is a tribute to Committee F–5 that it has brought together a diverse group of companies whose common thread of interest is in the making and using of business forms. In this group are business forms and business equipment manufacturers, naturally, but also included are paper manufacturers and suppliers of carbon black, dyes, waxes, chemicals, etc. Then there are the consumer representatives such as governmental agencies, testing laboratories, and insurance companies. It is from such companies as these and their technical personnel that the utility of our modern business copy products is derived. This publication, therefore, is dedicated to the business forms industry.

While those who have labored in this field for many years, and in particular, the members of Committee F–5, are very proud of their accomplishments as partially represented by this publication, we all realize that this is really just a beginning. We hopefully look forward to improved products, instruments, and understanding in the field of business forms in the years ahead.

D. M. Davidson
Applied research manager,
Standard Register Co.,
Dayton, Ohio 45401;
symposium chairman.

D. M. Davidson[1]

Legibility of Carbon Copy by an Instrumental Method

REFERENCE: Davidson, D. M., "**Legibility of Carbon Copy by an Instrumental Method,**" *Evaluation Methods for Business Form Copies, ASTM STP 482*, American Society for Testing and Materials, 1971, pp. 3–12.

ABSTRACT: Legibility of carbon copies has long been considered a subjective property where psychological factors play a significant part in the selection of acceptable quality. Some investigators have tried to place legibility on a more objective basis. The use of the scanning microdensitometer to make a profile of an individual image and then interpret that profile in numerical terms offers the most satisfactory approach to achieving that goal.

In this paper, a microdensitometer will be described, and the method that has been developed for using it in analyzing carbon copy image quality will be reported.

A mathematical model for the microdensitometer trace is proposed that will permit the interpretation of the effect due to character dimension and the correlation of parameters that may be used in describing legibility. The model also suggests several techniques that may be used in measuring these parameters.

Many of the variables affecting carbon copy quality include impressioning conditions, material differences, and form construction. Numerical data from the scanning microdensitometer technique will be used to illustrate the effect of some of these variables.

KEY WORDS: legibility, carbon paper, reproduction, forms (paper), microdensitometer, impact impression, manifolds, evaluation, tests

Everyone who can read considers himself an expert in judging legibility. The human eye is a fine optical instrument that can distinguish subtle differences between copies containing impressioned characters. However, there are deficiencies in human judgment and uncontrolled variables in selections made by eye that have made it desirable to have instrumental methods for measuring legibility of copy. Individuals tend to have esthetic preferences and biases that will produce considerable disagreement in choosing the best copy, particularly as the copy becomes more legible. Evaluation by eye requires a reference specimen which may not be readily obtainable or equivalent to previous reference specimens. Furthermore, it is very difficult to establish a numerical value that can be used for grading the specimen in

[1] Applied research manager, Standard Register Co., Dayton, Ohio 45401. Personal member ASTM.

question. There have been specially trained observers and panels judging legibility similar to those used for taste and odor evaluations. These groups tend to give improved agreement and even a type of numerical rating that is useful. However, the development and maintenance of such a panel is expensive and difficult. In addition, regardless of skill, an observer cannot judge satisfactorily from among a large number of specimens.

Lant[2] has described a statistical method of comparing carbon paper by observers. A matrix is completed by impressioning identical characters in successive steps. In the first step, approximately half of the spaces are filled randomly using one carbon. Then the second carbon is used to fill the remaining spaces. The observer picks a reasonable number of the "best" characters. From these choices, the relative superiority of one carbon over the other can be determined, using statistical techniques for calculations.

The first reported instrumental device for measuring legibility was that developed at Standard Register in 1951 [*1*].[3] This device was a microdensitometer that recorded a profile of the character's reflective density as a small area is scanned. The concept for judging the legibility was based on the difference between the peak reading and a base line defined by selecting points that correspond to the original width of the character.

A comprehensive report by Gartaganis [*2*] describes a method employing a microdensitometer. An equation was developed to correlate the instrumental results with a trained panel.

Current developments in optical scanning recognition of printed or impressioned characters emphasize the instrumental concept in judging legibility. The Kidder 081 test unit is used widely by printers to establish the numercial density of individual characters. Forsythe[4] has attempted to use the Kidder 081 as a scanning device that will yield a profile suitable for legibility interpretation.

A similar device has been described by Vaccaro [*3*] (U.S. Patent 3,460,143), and the concept of interpreting a density profile is discussed. More recently, Malone [*4*] reported a scanning method for evaluating electrostatic reproductions.

The group at Oxford Paper Company[5] also has developed an impressioning device and a scanning densitometer that gives a density profile of carbon images.

In this paper, the improvements in the instrumentation and method of evaluation of legibility as evolved at the Standard Register Company are described. In addition, a mathematical model is offered for the density profile which should permit better interpretation of variables than was previously possible.

[2] Lant, A. A., private communication.

[3] The italic numbers in brackets refer to the list of references appended to this paper.

[4] Forsythe, J. J., private communication.

[5] See p. 13.

Procedure

The instrument used in this study [5] is one that was designed and built at Standard Register Company. It is a scanning microdensitometer with the following features: It uses three tungsten bulbs to give uniform 45-deg illumination of the image area. A modified microscope stage serves to provide centering and movement for the image specimen. This stage is given in the scanning direction and can be adjusted manually in both directions. A lens system magnifies the image 20 times onto a ground glass plate. Magnification can be changed as desired by changing the lens system. The area to be scanned is obtained by mounting a 45-deg mirror on the ground glass plate. The dimensions of this mirror determine the cell size. This is also changeable. In this study, a rectangular scan spot 0.005 by 0.042 in. was selected. The mirror reflects to a photomultiplier tube whose readout is recorded on a chart.

Any impressioned image can be scanned to give a density profile. However, it is most convenient to use a zero or capital "O" produced by either a typewriter or computer printout. But these devices give variable characters that are not suitable for scientific comparisions of variables involving carbons or paper. A more precise test is obtained using an imaging device based on the pendulum principle. An image consisting of three sets of parallel lines has been selected. The dimensions of these lines are as follows: width 0.010 in., length 0.377 in., center-to-center distance between lines 0.040 in. This is slightly finer line than the normal impressioned character and about four times as long. It is much easier to center and scan than a normal zero, and reproducibility is considerably better.

Specimens are prepared by interleaving carbon and paper to give the number of parts needed for comparison. The most frequently used test set is one containing eight plies. A standard set is 8½ by 11 in. which can be inserted in a typewriter for test; it also can be punched with marginal holes and run through a computer printer; or it can be impressioned on the laboratory pendulum device. All methods may be used, but for comparisons the same impressioning method should be used. In practice, it is better to run all specimens to be compared in sequence and include a reference specimen whose characteristics are well known.

A sufficient number of characters are scanned to give a minimum of twelve readings. These are then averaged for reporting.

Discussion of Results

The density profile of an image contains a number of elements that should be considered. The most obvious element is the darkness or maximum density. This is an easily obtainable value and has considerable merit in describing an impressioned character. It is very rare that individuals will disagree on the relative rating of copy on the basis of darkness alone.

Another element is the background reflectance of the surrounding paper. This is also useful, and, while it is more difficult to discern, it is not likely to cause controversy among observers. For a simple character, such as described above, there is another element that can be defined. It is the valley that occurs between two maximum density peaks. Whenever the lines are close enough together that the profile does not return to the background level, this valley reading has a significance. Gartaganis [*2*] found that this value gives the best correlation with misreads from a panel of observers. This work was repeated and refined by Skees and Osmon [*6*] and the results confirmed. For this reason, this valley has been termed "legibility" and will be referred to as such in this report.

The application of this numerical method is very useful in examining the variables involved in carbon copy systems. The following examples demonstrate the effectiveness of this approach.

One important part of the business form copy system is the carrier tissue. The thickness, texture, resilience, and uniformity of the tissue has a major influence on the quality of copy obtainable from a formset. Table 1 shows the results of comparing the darkness and eligibility values for three different tissues with three different carbon formulas. All materials shown here are of acceptable commercial quality. However, it is easy to see that Tissue 1 is superior to Tissues 2 and 3. It should be noticed that in some instances the darkest copy is not the one with the highest legibility value.

TABLE 1—*Effect of tissue grade* (*4th copy of 8-part set*).

	Darkness			Legibility		
Formula	A	B	C	A	B	C
Tissue						
1	73.7	77.3	82.4	25.5	27.7	29.7
2	66.7	76.7	81.0	23.5	27.0	24.7
3	62.7	76.1	80.2	23.0	24.1	24.3

An obvious effect can be found for different ply levels in a formset (see Table 2). Darkness tends to drop rapidly for the first three plies in an eight part set but levels off there. Legibility drops more drastically and continues to drop throughout the set. This is a reflection of the continuing spreading of the image due to the paper thickness.

Carbon formulation has been a black art in every sense of the word. Each formulator has had his pet ideas and concepts, and each formula has been treated as a valuable secret. But in essence, the carbon ink used for one-time carbon is a fairly simple system. It is merely a mixture of vehicle and coloring matter. The coloring matter is generally carbon black, while the vehicle is a blend of hard wax, oil, and paraffin. Table 3 shows the darkness and legibility

TABLE 2—*Effect of ply (8-part set—pendulum impressions).*

Formula	Darkness A	Darkness B	Darkness C	Legibility A	Legibility B	Legibility C
Ply No.						
2	76.5	81.8	82.8	49.3	48.7	48.2
3	71.7	79.2	80.8	40.1	40.8	42.8
4	67.2	76.0	77.7	31.2	35.4	33.2
5	59.9	73.4	74.3	24.7	24.7	25.3
6	60.3	70.4	73.8	18.7	20.5	19.6
7	59.9	69.7	72.1	14.3	16.0	14.0
8	60.4	70.7	72.3	10.9	11.5	10.2

TABLE 3—*Effect of formula ingredients (8-part set—pendulum impressions).*

Formula	*x-1*	*x-2*	*x-3*	*y-1*	*y-2*	*y-3*
Hard Wax, %	10	10	10	15	15	15
Paraffin, %	25	20	15	20	15	10
Oil, %	35	40	45	35	40	45
Darkness, D:						
4th Copy	81.5	79.9	81.6	79.6	79.0	82.6
7th Copy	74.8	76.6	75.8	72.7	71.6	77.9
Legibility, L:						
4th Copy	38.2	35.9	37.8	39.1	39.1	35.6
7th Copy	22.5	20.0	22.1	21.3	20.9	20.1
Sharpness constant, K:						
4th Copy	1337	1285	1319	1405	1419	1225
7th Copy	870	751	839	850	848	740

values of six formulas where the vehicle system is varied. It should be pointed out that these results do not permit automatically the selection of the best product, since other properties such as smudge, ease of coating, and age stability are of equal importance to the overall performance of the carbon.

The impressioning method is also a variable of considerable importance. In Table 4, the relationship between the high-speed computer printer and the laboratory pendulum device is shown for five different one-time carbons [7]. The correlation is good for all except Carbon D. This is a handwrite or roller impression carbon and performs poorly when impacted, especially so when the blow is light and rapid, as with the computer printer. It is also of interest to note the variation that exists with the computer printer as compared to the laboratory pendulum device. Figures 1*a* and 1*b* show the images and the scan profiles for the same set impressioned on the computer printer and by the pendulum. The extreme variation in the darkness and legibility readings are much greater for the computer printer than for the pendulum. It is apparent that for fine differences to be discernible either a highly controlled impressioning device must be used or a large number of statistically selected specimens must be measured. Figure 2 is a photograph of a Standard Register Legibility Reflectance Meter Model 11.

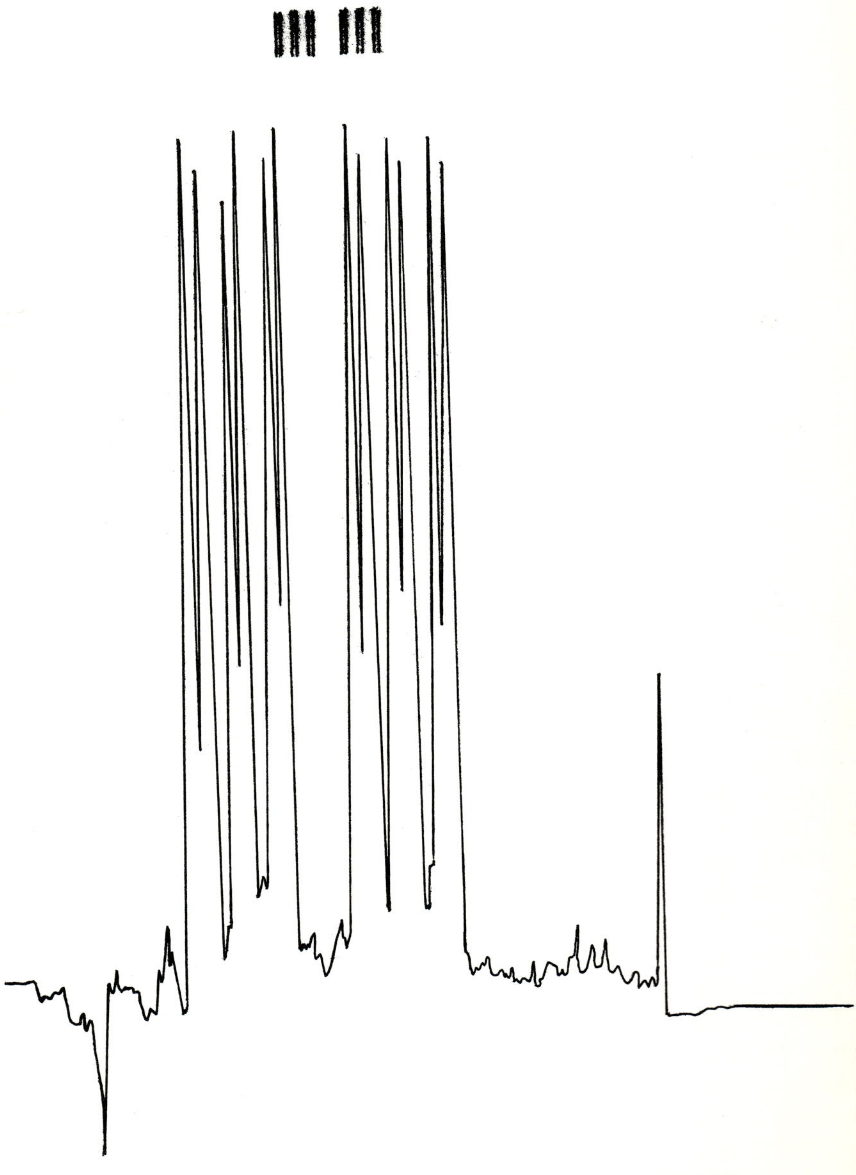

FIG. 1*a*—*Recorder trace of specimen produced by pendulum impact.*

TABLE 4—*Effect of impressioning method (4th copy of 8-part set).*

	Darkness		Legibility	
Formula	Pendulum	1403-2	Pendulum	1403-2
A.	53.5	50.4	27.1	27.4
B.	67.4	66.4	34.7	31.7
C.	73.8	71.3	36.7	39.5
D.	69.6	53.7	34.9	22.5
E.	71.9	70.9	41.5	35.6

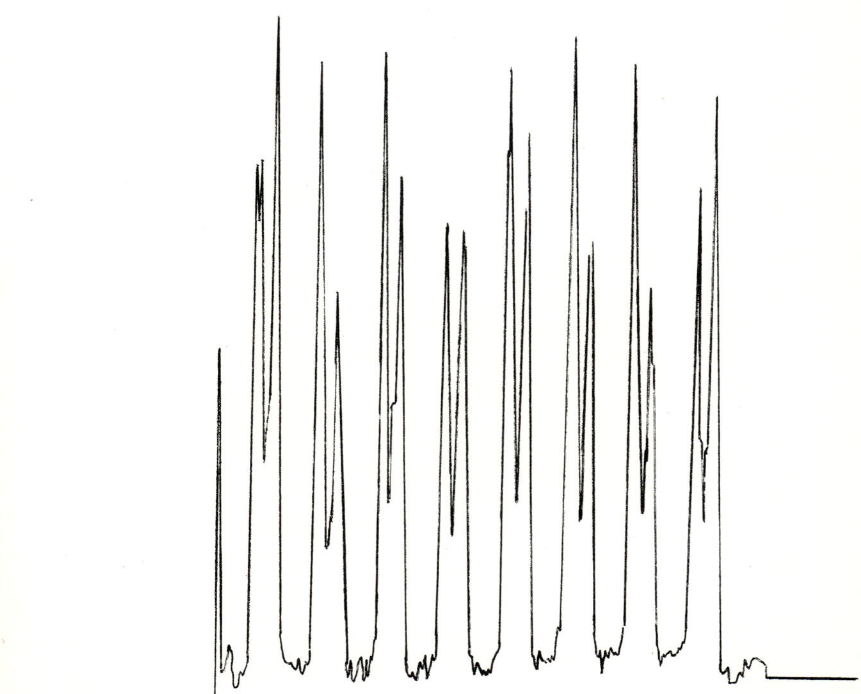

FIG. 1*b*—*Recorder trace of "O's" from high-speed printer.*

FIG. 2—*Standard Register Legibility Reflectance Meter Model 11.*

Mathematical Model

Frequently, it is of value to establish a mathematical model for an observable phenomenon in order to facilitate the interpretation and perhaps to improve the techniques for measurement [*8*]. In this case, the density profile of an impressioned image closely resembles a normal distribution curve which can be represented by the equation (Booth [*9*]):

$$y = ae^{-kx^2} \qquad (1)$$

where:

y = density (or darkness) at any point x
a = maximum density (or darkness), and
k = a constant for the system.

If two lines are displaced by the distance d, the peak and valley for the profile can be calculated from the equation:

$$\text{peak} = P = a \qquad (2)$$

$$\text{valley} = V = ae^{-\frac{kd^2}{4}} = Pe^{-\frac{kd^2}{4}} \qquad (3)$$

Since legibility $= L = (P - V)$

$$L = P(1 - e^{-\frac{kd^2}{4}}) \qquad (4)$$

and it is seen that legibility depends on maximum density and the distance between the lines forming the character.

In our measurement the peak value is not equivalent to darkness but is related to it as follows:

$$D = P + B \qquad (5)$$

where P is the peak reading and B is the paper background reading.

Therefore, a correct relation between the units that have been selected is:

$$L = (D - B)(1 - e^{-\frac{kd^2}{4}}) \qquad (6)$$

It is possible to establish a value for L without having a pair of lines to scan. Any single line will give a characteristic profile that fits the proposed equation. Therefore, all that is necessary is that the maximum density, paper background, and density at some known point be measured; then L can be calculated. A simple technique would be to find the point (X) where density is equal to maximum density time $1/e$.
Then:

$$ae^{-1} = ae^{-kx^2} \qquad (7)$$

or

$$k = x^{1/2} \qquad (8)$$

and then L can be calculated from Eq 6 by using a standard value for d.

Another idea would be to make a character that contained a series of lines that fall sequentially closer together. Whenever the lines become indistinguishable, the legibility L can be determined, since at that point, a plot of valley readings logarithm against the distance squared will yield a straight line with the slope of $k/4$:

$$V = ae^{-\frac{kd^2}{4}} \qquad (3)$$

$$\text{Log } V = a - (k/4)d^2 \qquad (9)$$

In practice, the rating of copy can be made by simply assigning a number to each distance and using that number as the rating whenever that is the largest gap that becomes indistinguishable.

Finally, it may be asked why not utilize the constant k more in the comparisons of copy. It can be shown that k is related to the ratio of L/P which is sometimes referred to as "sharpness." This is a useful value, but, as noted previously, legibility (L) is believed to be more pertinent in evaluating copy.

Acknowledgments

The author thanks the Standard Register Company for permission to publish this paper. Many members of the Research Department contributed to the work reported here. In particular, credit should go to Hugh Skees, Ernest Emmert, James Martin, Charles Adkinson, Jack McClannan, Stan Kebo, Phil Osman, and Lydia Zimmerman. Additional acknowledgment is accorded W. T. Paulson, research technical director of Standard Register, for his interest and encouragement of this work.

References

[1] Davidson, J. T., *The Office*, June 1952, pp. 63, 88.
[2] Gartaganis, P. A., *American Ink Maker*, Sept. 1963, pp. 25–29.
[3] Vaccaro, A., U. S. Patent 3,460,143, 5 Aug. 1969.
[4] Malone, D., paper presented at TAPPI Graphic Arts Conference, Oct. 1969.
[5] Skees, H. B. and Adkinson, C. A., Standard Register Engineering Report, 1965.
[6] Skees, H. B. and Osman, P., Standard Register Engineering Report, 1964.
[7] Martin, J. R., Standard Register Engineering Report, 1969.
[8] Davidson, D. M., Standard Register Engineering Report, 1969.
[9] Booth, A. D., *Journal of Scientific Instruments*, Sept. 1969, pp. 822–823.

R. H. Graham,[1] *T. A. Fadner,*[1] *and W. T. Barker*[1]

Laboratory Evaluation of Manifolding Paper Performance

REFERENCE: Graham, R. H., Fadner, T. A., and Barker, W. T., "**Laboratory Evaluation of Manifolding Paper Performance,**" *Evaluation Methods for Business Form Copies, ASTM STP 482*, American Society for Testing and Materials, 1971, pp. 13–22.

ABSTRACT: A method for laboratory evaluation of manifolding papers is presented. An impact imaging device and an electro-optical image evaluator are described. With these instruments, field performance of commercial manifolding products can be predicted. In addition, test methods have been developed which characterize energy transfer, image transfer, and image receptivity, three primary product properties that affect manifolding. The use of the instruments and methods as an aid in the development of high manifolding products is illustrated.

KEY WORDS: manifolds, impact printers, image, carbonless papers, bond paper, carbon paper, evaluation, tests

A means of accurate and rapid evaluation of image quality is needed in the development of manifolding paper for impact printers. This requires three things:

1. A reproducible impact imaging test device.
2. An objective evaluation of image quality.
3. A correlation of laboratory instrumental results with manifolding performance on impact printers in the field.

Oxford has developed a reliable impact imaging device and an image evaluation instrument. We have shown that laboratory results can be correlated with field performance. Specifically, we have identified three product properties that affect manifolding. These are:

1. *Energy Transfer*—This is the ability to transfer impact force through a stack of paper sheets.

2. *Image Transfer*—This is the ability of the product to release the backside coating or chemicals upon impact, and transfer them to the next sheet in the stack.

3. *Image Receptivity*—This is the ability of the product to receive the transferred image, on its front side.

[1] Research and Development Department, Oxford Paper Company, Division of Ethyl Corp., Rumford, Me. 04276.

The following sections describe the instruments, show how they can be used to predict field performance, and illustrate laboratory quantitative measurements of individual factors affecting manifolding in pressure transfer systems.

Impact Imaging Device

The impact imaging device is illustrated in Fig. 1. It consists of a specimen holder, an image key, and a pendulum type hammer. Impact imaging is accomplished by raising the hammer to a specific height and allowing it to fall and strike the specimen positioned on the specimen holder in front of the imaging key. The impact pressure can be varied over a wide range to simulate different printers by merely changing the drop angle of the pendulum. A series of images are formed by rotating the specimen.

Special three-bar keys designed to form images suitable for scanning are used. Typical images made by the instrument are shown enlarged in Fig. 2. Three groups each of good and bad three-bar images are shown.

Image Evaluator

The image evaluator is illustrated schematically in Fig. 3. It is basically a scanning reflection microdensitometer. It consists of a microscope type objective, a source of illumination, a slit backed by a photocell, a recording photometer, and a rotating drum type specimen holder which permits rapid scanning of multiple images. Reflectance standards are built into the drum for calibration purposes. A beam splitter is used to permit focusing and visual alignment of the specimen.

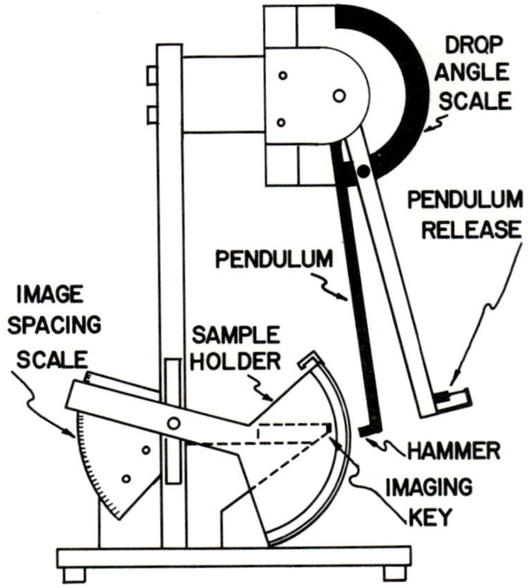

FIG. 1—*Oxford impact testing device.*

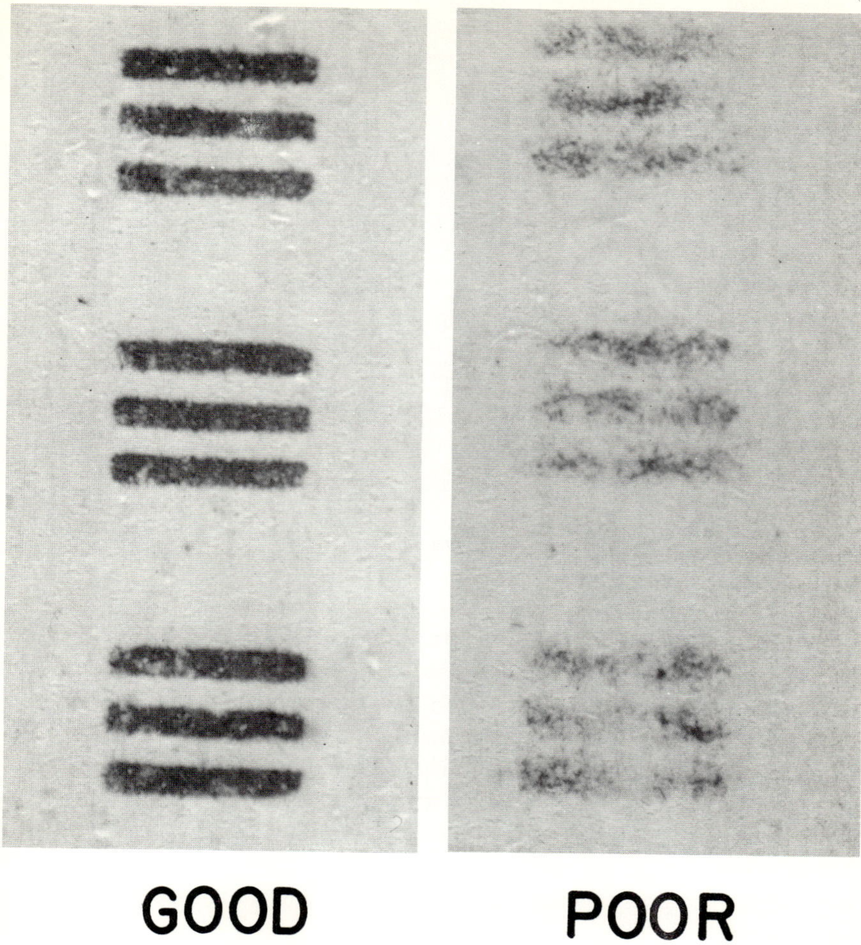

FIG. 2—*Typical test images (enlarged).*

The recording photometer and source lights are commercially available units. The optical system was assembled by Gaertner Scientific Corporation to Oxford specifications using a modified parfocal microscope to provide the beam splitter. The microscope eyepiece was replaced by an adjustable slit attached directly to a photodetector. The length and width of the slit can be adjusted to accommodate various image dimensions.

Mr. Davidson in his paper mentions the need for multiple specimens to obtain reliable measurement.[2] Our own work shows that a large number of images per specimen must be scanned to obtain a statistically valid measure of the product.[3] Therefore, our scanning unit was designed to hold an 11-in. paper strip which contains up to 100 image characters.

[2] See p. 3.
[3] DeFelice, V. N. and Stewart, J. E., private communications.

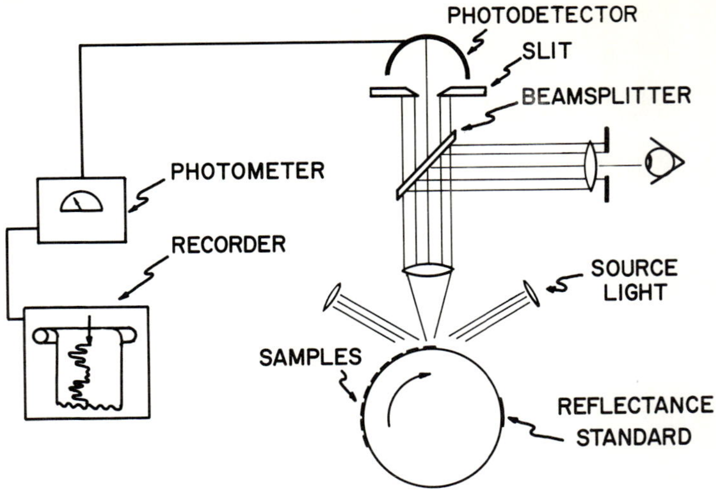

FIG. 3—*Schematic, image evaluator.*

As the drum rotates, the characters are passed beneath the optical system which projects a magnified image onto the adjustable slit. A photodetector directly behind the slit generates an electrical signal proportional to the light reflected from the paper specimen. This signal is recorded on a chart recorder. Figure 4 is a photograph of our actual equipment.

Literature surveys and our laboratory studies have shown that image contrast is the most useful measurement for predicting image legibility.[4–6] Therefore, the image evaluator was designed to measure the contrast of black images on nonglossy white papers. Most manifolding papers meet this requirement.

Let us refer again to Fig. 2. It was difficult to obtain a measurement of overall legibility of the relatively nonuniform and sometimes poorly defined images characteristic of impact manifolding systems. Since legibility is dependent on overall image density, we adjusted the slit size to optically average a large portion of the image. Using this large slit, each single bar image will be recorded as a single peak.

A typical trace of several three-bar images is shown in Fig. 5. The peak value D_1 corresponds to the optically averaged image density. The base line D_2 corresponds to the light reflected from the unimaged background area. The difference between average peak density and average background density $(D_1 - D_2)$ is used as our measure of contrast.

[4] Gartaganis, P. A., "Legibility of Carbon Paper Prints," *American Ink Maker*, Vol. 41, No. 9, Sept. 1963.

[5] Davidson, J. T., "Measuring a Forms Legibility," *The Office*, June 1952.

[6] Oxford Paper Company Research Reports, 1966–1969.

FIG. 4—*Oxford image evaluator.*

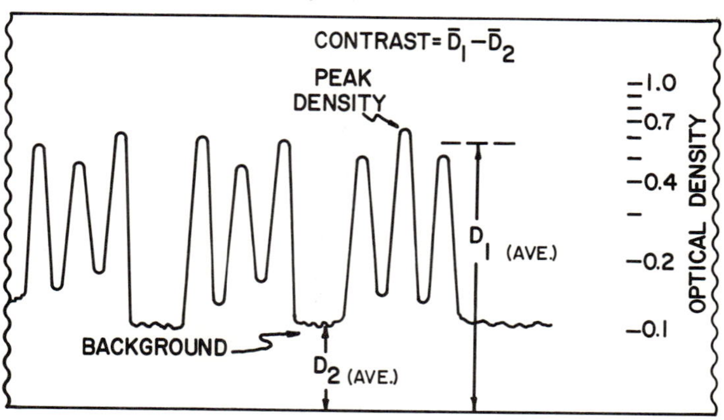

FIG. 5—*Typical image evaluator chart record.*

Use to Predict Field Performance

We have shown that this instrument may be used as an in-house test to predict field performance with manifolding papers.[3] A series of ten and twelve part books of carbonless papers of different quality levels were prepared and imaged on an IBM 1403, N1 printer. A trained panel of observers rated the manifolding ability of these papers in terms of the last acceptable

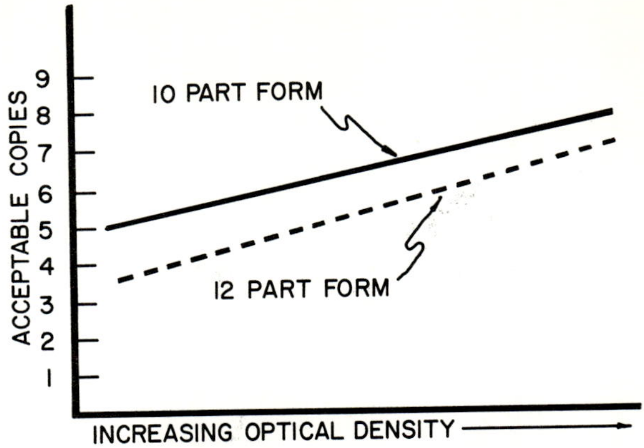

FIG. 6—*Manifolding versus optical density.*

part in the set. Specimens of the same materials were imaged on the impact tester and evaluated with the image evaluator. The correlation between manifolding ability, as judged by the panel, and the average densities obtained on the image evaluator are shown in Fig. 6. The correlation in this case was significant to the 99 percent confidence level, but more work will be required before this can be used as a routine control test.

Use as a Development Tool

The image evaluation equipment developed by Oxford can be used to evaluate manifolding products in terms of the energy transfer, image receptivity, and image transfer: three basic characteristics which influence manifolding.

Energy Transfer

Energy transfer is a measure of the ability of the paper to transmit printer impact effectively through multiple parts of manifolding paper forms. This concept has been described previously by Lindgren.[7] A stack of the paper to be tested is prepared in which the last two parts are a standard transfer and a standard receptor sheet, respectively. The specimens are imaged on the impact tester at a 55-deg drop angle, and the image density on the standard receptor is measured. Energy transfer is expressed as the average optical density of the image formed on the standard receptor sheet. We will illustrate the use of the energy transfer concept in a later figure.

Image Transfer—Image Receptivity

The ability of the materials to transfer and receive the image forming components upon impact is called image transfer and image receptivity,

[7] Lindgren, R. M., "Physical Properties of Paper That Affect Manifolding," paper presented at the 25–29 May 1969 Tappi Coating Conference, Minneapolis, Minn.

respectively. These properties are measured using a two part set; for example, a carbonless CB and CF set or a bond/carbon/bond set. A two part set is used to eliminate the energy transfer effects of a multipart set. Figure 7 is a plot of image density as a function of pendulum drop angle for a typical manifolding paper system. This type of plot is particularly convenient in comparing the component performance characteristics of different products. By proper makeup of the specimen, the image transfer efficiency of a CB-coating or the receptivity of a CF-coating can be evaluated independently of each other.

Figure 8 shows the image density versus pressure relationship for two different carbonless products. The products were imaged first against their own receptor sheet and then against a sheet of ordinary register bond. Product E obviously has the best transfer with its own receptor, Curve E_{CF}; and the least transfer to the bond sheet shown as Curve E_B. Product X shows little difference in manifolding capability between its own receptor and bond paper shown as Curves X_{CF} and X_B, respectively. Product E has good specificity; Product X does not.

Figure 9 is a plot of image density versus pendulum drop angle or impact pressure for three experimental carbonless transfer coatings evaluated against a standard receptor sheet. It is apparent from these curves that Coating A, is less sensitive to changes in impact presssure than Coatings B or C. Coating A would have high manifolding capability, but would be a high scuff product, since it is susceptible to transfer at low pressures.

The major goal in the development of carbonless papers is to make a carbonless paper system that will yield a dense clearly defined image on impact in the printer. However, that image should be resistant to scuff or

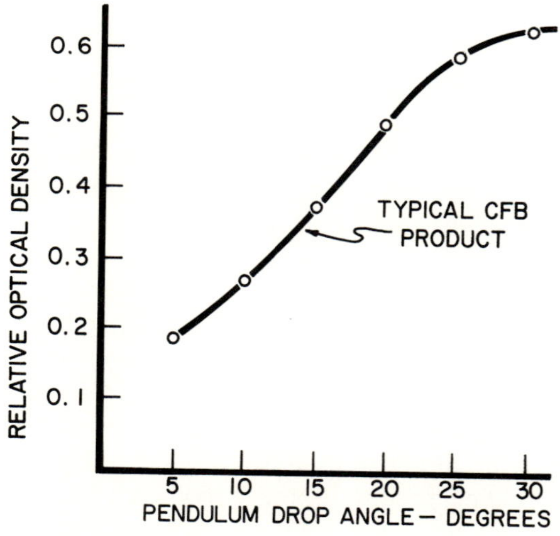

FIG. 7—*Optical density versus drop angle.*

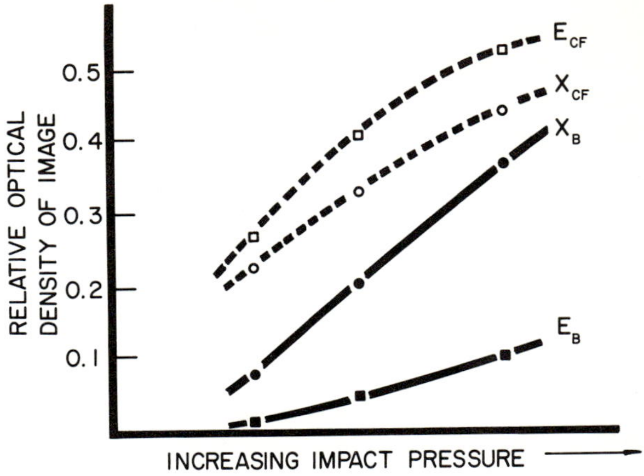

FIG. 8—*Image density versus impact pressure.*

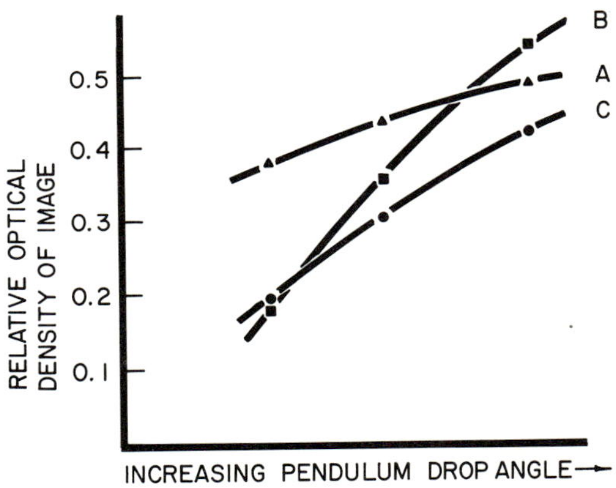

FIG. 9—*Image density versus drop angle for experimental transfer coatings.*

setoff in normal handling. This means that below the low impact pressures encountered in the printer, the image transfer should be minimal, but above this pressure, maximum transfer is desired. This concept is illustrated in Fig. 10.

Tests have shown that a 15 to 20-deg drop angle on the impact tester gves image densities similar to those obtained in a multiple book on an IBM 1403 N1 printer. This is represented by the shaded area in Fig. 10. An ideal product would exhibit little or no transfer below a 15-deg drop angle corresponding to the minimum impact pressure of the printer and would exhibit maximum transfer at all pressures above this value. Our presentation of the ideal high manifolding product is the heavy dark line. A typical imaging carbonless paper found in the field is represented by the dashed line. A product which gives high optical density at low impact pressures or drop

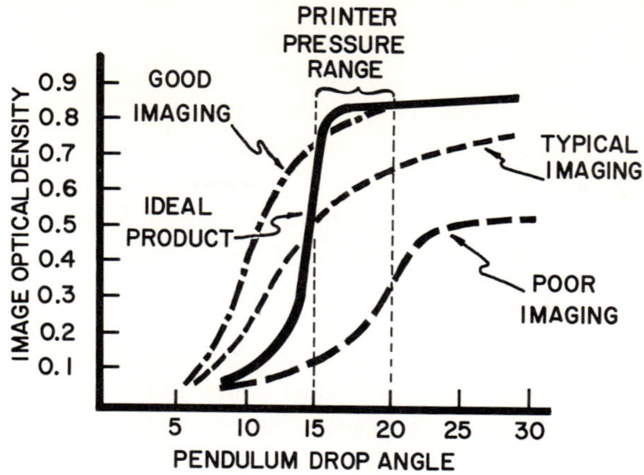

FIG. 10—*Imaging performances of physical transfer products.*

angles will have a high scuffing tendency. The good imaging curve shows the image density versus pendulum drop angle characteristics for a high manifolding, high scuff product. The poor imaging curve shows the response characteristics for a low manifolding, low scuff product. This product, incidentally, has been shown to be unacceptable for high manifolding applications.

We would like to give you an example of how we have used this image evaluation equipment as an aid in the development of high manifolding register bond paper. A series of paper machine experiments were made to determine the effects of machine and furnish adjustments on manifolding performance. Two of these have been selected for this example (we refer to these as Trials A and B). Figure 11 shows the relative performance characteristics of these trial papers compared with a standard high manifolding bond. The panel's rating of manifolding performance of the trial papers and of the standard product are shown in the first line in Fig. 11. Energy transfer is shown in line 2. Image receptivity at two impact levels is shown in line 3.

	TRIAL A	TRIAL B	STANDARD
MANIFOLDING ACCEPTABLE COPIES	4.30	5.70	6.00
ENERGY TRANSFER	0.63	0.80	0.81
IMAGE RECEPTIVITY (15°/20° DROP ANGLE)	0.40/0.50	0.37/0.51	0.46/0.61

FIG. 11—*Image evaluator data for experimental papers.*

Note that paper made in Trial A has both poorer image receptivity and much poorer energy transfer than the standard. Note that the energy transfer of the paper made in Trial B is approximately the same as the standard, but that its image receptivity was lower. It was our conclusion that desired improvements in manifolding performance of Trial B could be made by improving the receptive properties of its surface. Improvements in Trial A would require improvements in energy transfer and image receptivity.

Conclusions

In summary, we have described an impact image test device and an image evaluator. We have shown that the image evaluation equipment can be used to predict the field performance of manifolding papers. We also have shown that these instruments can be used as development tools to evaluate energy transfer, image transfer, and image receptivity; basic factors affecting manifolding. It is believed that use of these instruments can lead to the development of improved manifolding papers.

Acknowledgments

The authors thank the Oxford Paper Company for permission to give this paper. Dr. K. V. Kraske was instrumental in the early development of the test devices. V. N. Defelice and J. E. Stewart were particularly helpful in statistical evaluation of the data.

R. G. Reynolds[1]

Instrument Evaluation of Electrostatic Paper

REFERENCE: Reynolds, R. G., "**Instrument Evaluation of Electrostatic Paper,**" *Evaluation Methods for Business Form Copies, ASTM STP 482*, American Society for Testing and Materials, 1971, pp. 23–35.

ABSTRACT: This paper is a report on research to find a means of analysis of ZNO coated light sensitive paper from small specimens and be able to predict resultant copy. Present instruments can measure and record similar parameters, but predictable results could not be made since the range, level, and cycle of adjustments were not related to machine cycle. When the methods and controls presented are followed, it is possible to predict optimum copy and make estimates on latitude, background, loss of density of copy, as well as other characteristics related to machine conditions. None of the instruments are new nor are any of the parameters measured new. The research was in determining type and range of instruments, methods of operation, and calibration to duplicate machine cycle. It was also necessary to determine and measure all elements of the machine cycle.

The first study is of the machine to determine light level and variations due to system and voltage. This includes light on both background and image. Other machine factors are exposure time, dark decay time, charge, time cycle of copy, and any special machine components such as preexposure, double exposure, etc. The combination of instruments then may be set up to duplicate these machine functions. These may include turntable, light source, electrometer, recorder, voltage generator, variac, and programmer to operate and control the instruments over all ranges and in same sequence of machine operation.

The third and of equal importance are methods and proof of calibrations. For example, a lightmeter reading on level of light source, even when taken by placing meter in specimen exposure position, is not enough. The calibration must prove that the light level is not an average reading or that there are no high or low points in the specimen exposure plane.

The entire system must be automatic and adjustable, equal to full range of machine adjustments. Production of the electrophotographic curves should take no longer than a similar copy in the machine. A test of the system is when you can make a copy on the machine, set the instruments at same levels, and then show in the recording the background, best copy, image loss or any problem or area that requires explanation.

KEY WORDS: dark decay, delta, corona, electrophotographic curves, electrostatic paper, foot candle, foot candle seconds, light dark decay, light decay, probes, evaluation, tests

Thirty-six years have been devoted by the author to the business machine field with twenty-nine years at Burlington dealing with supply products.

[1] Quality control manager, Olivetti Corporation of America, Burlington, N.J. 08016. Personal member ASTM.

Subjective evaluation had been the only test for supply products; therefore, we have developed attribute measurements, functional tests, and new instrument evaluation for them. This included in process, final test, development of new products, and evaluation of experimental products.

Complete quality control procedures have been established and recognized by vendors and buyers in the industry. The tests methods and instruments have been copied in many cases and have become part of standards or specifications.

Work in the electrophotographic field started in 1964 with the responsibility for establishing quality control procedures and methods of evaluation of electrostatic paper and toner. Previous methods of relating test procedures to functional use were very helpful. Instruments for color control and density evaluation were adaptable with new procedures to this new field.

There are and have been methods using instruments to charge electrostatic ZNO coated paper, measure the charge retained, expose to light, and measure the resultant light decay and in some cases dark decay.

This report is on work done to expand and combine instruments to measure all parameters including charge, second by second light decay, true dark decay, light dark decay (light dark decay, LDD, the reduction of charge on image area of charged ZNO coated sheet due to reflected light on the image), residual and double exposure, all related to machine conditions, and the calibration of instruments so that the electrophotographic curves will provide information that may be used as one of the techniques to predict copy from a roll type machine. The capability to predict copy from small specimens permits evaluation prior to large costly production of rolls and prior to commitment of large shipments.

The system has been used for quality control and evaluation of experiments, as well as an aide in evaluation of unidentified specimens.

The principle of this test is to determine every function of the copying machine, duplicate these functions in instruments, and calibrate all instruments.

Machine

Therefore, the first study is made of the copying machine to determine all factors that affect the copy produced.

The complete copy cycle must be timed and each variable measured. While these will vary with each machine, the order of occurrence that may change the basic factors are:

1. Charge—Determine: (*a*) Speed of paper, flow or movement of corona. (*b*) Is this a single or double corona. (*c*) What is distance from corona to the paper? (*d*) Any other relevent factors related to the charge.

2. Dark decay time—Is there dark decay time between charge and exposure? Is there dark decay time after exposure and before development?

3. What is speed of paper feed? Paper feed and machine cycle pattern will determine dark decay time and in some cases may vary dark decay time with different length of copy.

4. Light level or light intensity at point or area copy is produced. This must be identified and measured to 0.1 foot candle (FC). Uniformity of illumination must be known at every point of the exposure plane.

The variation of light level or light intensity must be known. An average light level cannot be used as the copy does not see an average. The machine sees point to point. Where the light intensity is highest, the copy may be light and broken, and, where the light intensity is low, the background will be dark. A second consideration of light level is how much does the level change with variation in line voltage and how much variation must be considered due to location and machine condition? The third light level consideration is the reflected light on the image. This will be referred to as light dark decay (LDD). This will vary with the width of the line, but lightmeter measurements show that broad areas receive about 10 percent of the light that the background sees, and the percentage increases as the line gets thinner.

For example, with a given background light level of 10 FC, the light level on the broad area would be 1 FC. As the width of the line decreases, the light on the line increases. Recognition of this will help to explain image loss when exposure is increased to eliminate background due to uneven light distribution.

5. Exposure time—How long is the paper exposed to a known light level? What is the increment of adjustment of exposure time? What is the total range of adjustment?

6. Determine the complete time cycle of the machine that is required to produce a copy. Example, the copy is produced by (*a*) feed, [*x*] feet per second; (*b*) charge, [*x*] seconds; (*c*) expose, [*x*] seconds; (*d*) feed, [*x*] feet per second; and (*e*) development, [*x*] seconds.

7. Determine development of foot candle seconds (FCS) required to produce optimum copy and variation. Is light intensity changed or is speed of paper feed changed?

8. The machine also must be examined for and consideration given to any special machine conditions required to produce copy. Examples might be preexposure or double exposure.

Instruments

The method of operation of the machine to produce a copy provides the information required to set up the proper combination of instruments to duplicate the machine cycle.

These will include:

1. A type of turntable and paper holder that will duplicate the method by which the machine holds the paper—this may be a back plate or belt combined with a method of charge and level from the same type of corona at the

same spacing as in the machine. The speed of paper flow and sequence of paper movement and charge must be similar to the machine cycle.

2. Light source—This must be of the same type and intensity, and capable of spanning the same range of variables and adjustments as in the machine. The light also must reach the specimen by the same method as in the machine. Example, light in the machine may reach the copy by off-target reflection or by passing through the target, etc.

3. A vital component of the set of instruments is an electrometer or similar instrument to detect the calibration charge (see 5) and changes in charge and residual on the paper being tested.

4. A recorder to record this charge and changes—Its speed and sensitivity must be greater than the rate of change in the paper. The recorder also must be equipped with an event marker to show when light is turned on to expose specimen and record when light goes off.

5. Voltage generator with range of 0 to 600 V—This is connected into the instrument system so that a calibrated voltage of the same polarity, equal to the nominal charge acceptance of paper, is placed on the same paper-holder area and recorded in the same way as charge on the paper.

6. Programmer—this has to be made (we suggest solid state) and consists of a series of timers or switches that will automatically control all parts of system through sequence of events equal to copy cycle of the machine. The programmer should have a range of adjustment equal to and greater than the range in the machine.

A normal cycle would require the operator to place the specimen in position and press the start switch of the programmer. The programmer would than:

1. Start turntable and recorder.
2. Turn on the corona.
3. Turn off the corona.
4. Stop turntable under probe (if paper stops in normal machine cycle).
5. Control dark decay time (duration adjustable).
6. Turn on light (duration adjustable).
7. Turn off light.
8. Control dark decay time (duration adjustable).
9. Turn on turntable and move to home position.
10. Turn off recorder feed.

The programmer also should have manual controls and bypass switches to permit calibration and use of instruments for sequence of events other than normal for the machine.

The programmer makes possible repeating the same test on a series of papers and aides in accurate adjustment of the cycle required for dark adapted light sensitive specimens under dark room conditions.

Calibration

If we consider knowledge of the machine and its cycle as one side of an equilateral triangle and the instruments as the second side, then the third side becomes calibration.

1. Calibration of light—The same method and lightmeter used to measure light on the paper in the machine should be used to measure light on the specimen. This will calibrate and permit adjustment of light intensity to 0.1 FC.

2. Calibration of light level—We refer here to the determination as to how even light intensity is to all areas of the specimen. If light is less in one area than in another then the light decay will not be the same, and the electrometer will read an unknown residual or average residual. Since toner is very sensitive to residual voltage, as can be seen in the copy from a machine with uneven light distribution, a specimen exposed to light in instruments is run through a developer section. If the light distribution of instruments is uneven, the developed specimen will show light and dark areas. If the calibration of light level is correct, the developed density of the specimen will be even.

3. The electrometer and recorder are calibrated and checked out with the voltage generator. A series of known voltages from 0 to 400 are placed on the same plane and position as the test specimen and picked up by the same probe which reads the specimen.

4. Since both high voltage of charge and low voltage of residual are important and recorder tape line width, line space, and trace line width reduce accuracy below 20 V and may not read below 10-V intervals, expansion and calibration of the recorder sensitivity at low voltages should be considered.

5. Event marker—Check to be sure events are recorded on tape. The most important is light, when the light is turned on the specimen and when it is turned off.

Operation and Interpretation

The instruments set to match the machine cycle and conditions are set up and calibrated in an ASTM Faraday room and operated under dark-room conditions.

The required number of specimens are cut from a dark adapted sample. The electrophotographic curves are produced to represent the complete range of machine copy settings.

These curves show graphic record of charge, dark decay, light decay for each half second, and residual due to level of light on the background. The series of curves will show, in addition to the results from all machine settings from lightest to darkest adjustment, an expansion of the residual for accurate background interpretation. The light level in the instruments is changed to match that of the light on the copy image. The curves that are produced will show residual in relation to image. Since the production of curves by instruments duplicates the machine cycle, the curves will show the change in copy due to changes in machine adjustment or what would happen to copy if subjected to a second charge and a second light cycle.

Examination of actual curves will show each of the points discussed and how conclusions are drawn and predictions made.

Figure 1 shows machine setting of two for very dark copy from normal exposure on Paper A. The copy produced would have a very dark background and low contrast.

Point A = charge acceptance, 360 V;
Points A to B = dark decay;
Point B = light on and exposure starts;
Points B and C = total exposure;
Point C = light off and exposure ends;
Points C and D = dark decay;
Point D = residual voltage that represents background of copy; and
Points A' through D' are the second complete cycle on the same specimen and made to determine specimen reaction to existing machine conditions.

During first exposure of 1.4 s, the light decay did not start as soon as the light came on; decay was slow and did not stop after the light went off. The resultant residual was 265 V. During the second exposure of 1.4 s, the light decay shows usual slow start, increased speed, much greater loss after the

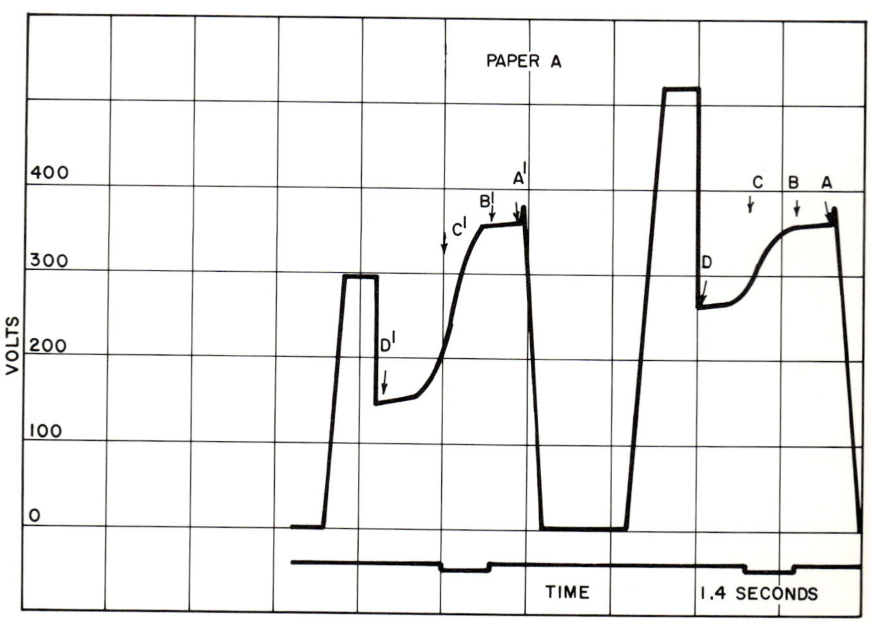

FIG. 1—*Machine setting of two for very dark copy from normal exposure on Paper A.*

light went off, and resulted in residual voltage of 145 V. The copy will show high background and poor contrast, and the portion of the copy that takes two cycles will show less background and better contrast.

Figure 2 shows the same machine at 2.1-s exposure on Paper A. Background is reduced, but the second cycle still shows less background.

Figure 3 shows 2.8-s exposure on Paper A. The background of both are about equal, and are reduced to show that copy from some targets may be expected.

Figure 4 shows 3.5-s exposure on Paper A. Normal optimum machine setting is used for the best copy with this paper. The light goes off at a point where there is no continued decay, and residual is at normal level for the paper.

Figure 5 shows 4.2-s exposure on Paper A. *Note*—Point *D* of Fig. 5, extended exposure has not reduced residual; therefore, we may expect constant residual at this level. Point *D* is an expansion of residual by increased recorder sensitivity to permit positive identification.

Note also that at Point *D* decay has ended with the light still on. This indicates two areas, (1) optimum exposure is less than 4.2-s duration and (2) extended exposure, longer than required to reduce background to constant residual only will reduce or burn out the image of the copy (Fig. 6).

The same machine cycle with Paper A represented but with extended dark decay for positive identification and light level set in the instruments at

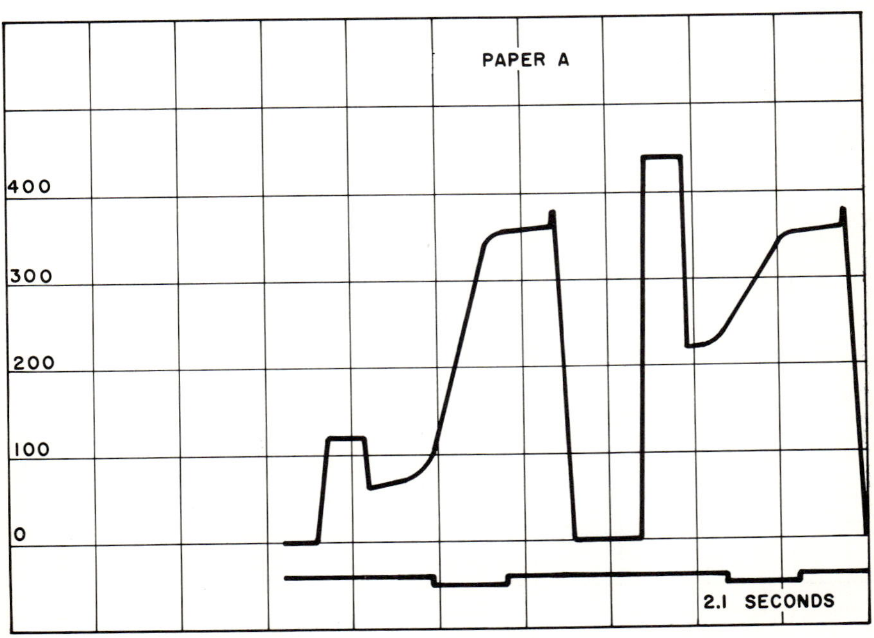

FIG. 2—*Same machine at 2.1-s exposure on Paper A.*

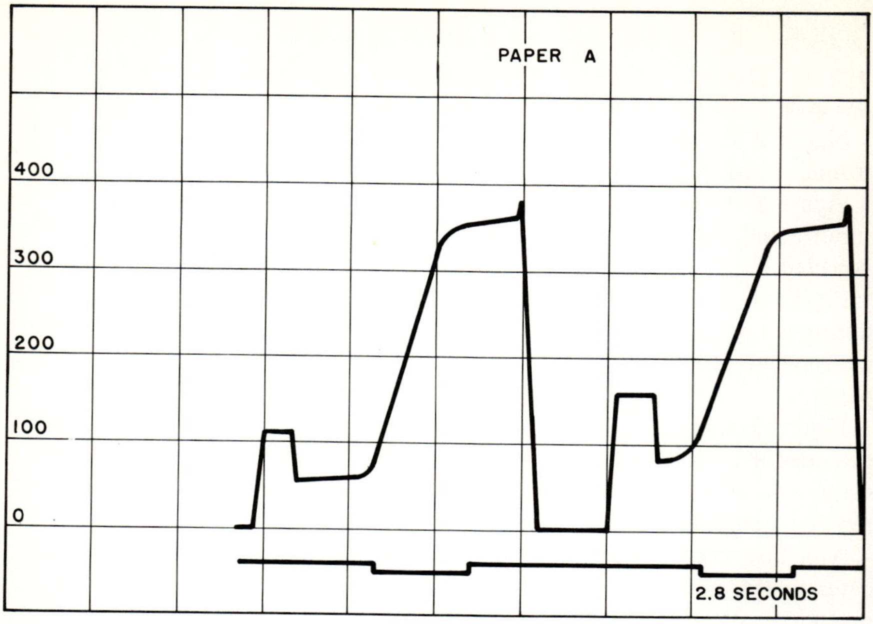

FIG. 3—*Same machine at 2.8-s exposure on Paper A.*

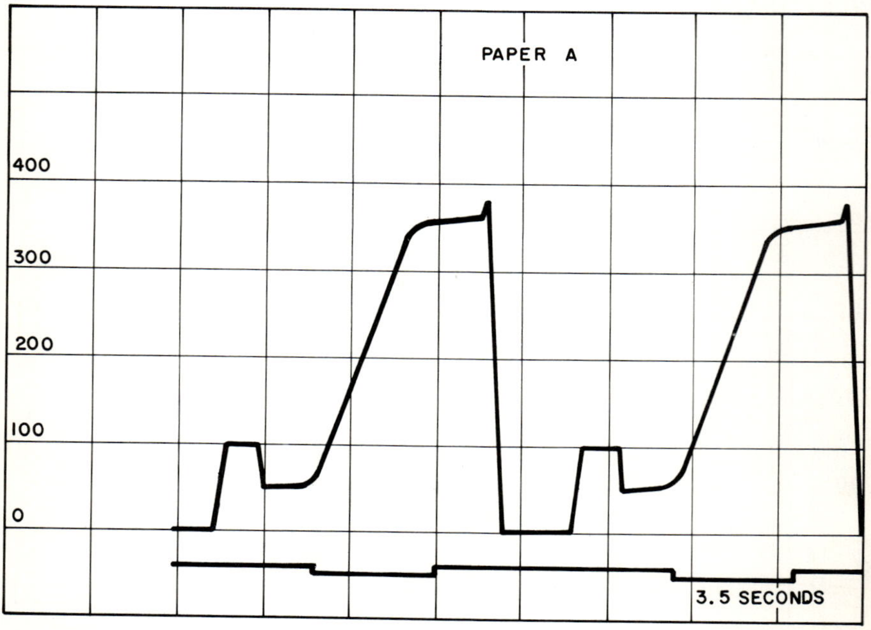

FIG. 4—*Same machine at 3.5-s exposure on Paper A.*

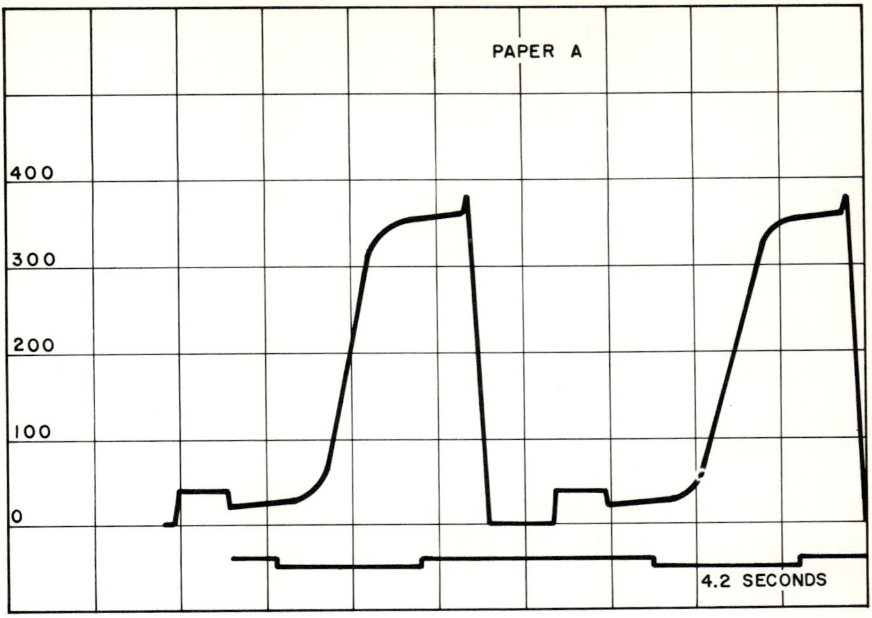

FIG. 5—*Same machine at 4.2-s exposure on Paper A.*

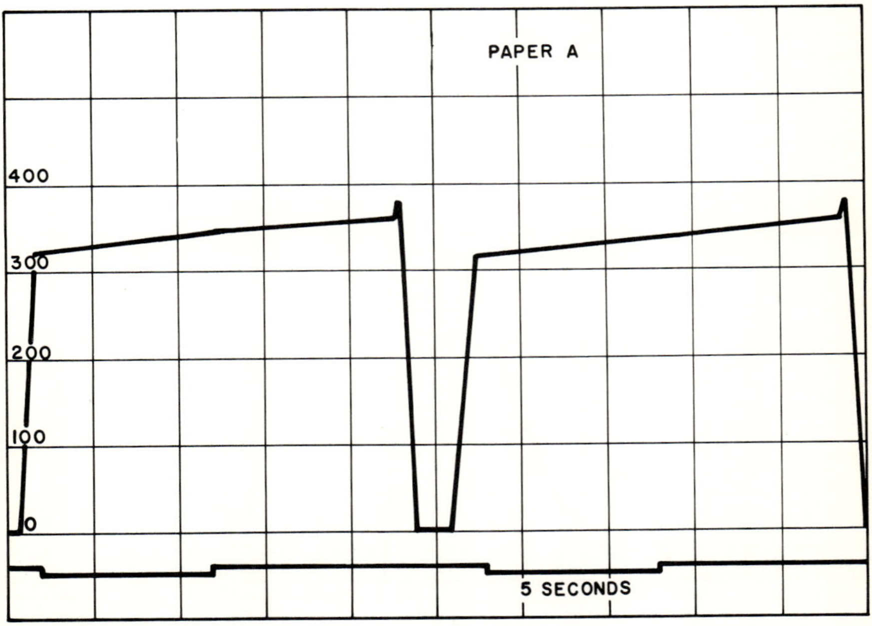

FIG. 6—*Same machine cycle on Paper A but with extended dark decay and light level set in the instruments.*

intensity of light on the broad area of image is shown in Fig. 6. This represents the LDD and shows decay of the image on the copy. Extended exposure would reduce the image, and line copy in the area of the second cycle would be reduced more than in the one cycle area. Some papers show rapid loss under this test.

Figure 7 shows that the 3.5-s exposure is the same as in Fig. 4 but with different paper, Paper B. The charge is lower, and the residual is 50 percent higher with the result that delta (delta is the term used to express net charge remaining to produce copy after dark decay and light exposure) is less. We may expect optimum exposure to be six or seven instead of four as with Paper A used in Fig. 4 under the same target conditions.

A third type of paper, Paper C, at short exposure is shown in Fig. 8. The higher background is present with the same reduced background resulting from the second cycle.

Paper C at extended exposure is illustrated in Fig. 9. *Note* 1—The constant residual and duration of exposure required shows this to be a slow paper. From knowledge of machine reaction to normal voltage variations and various targets, we may predict that with some targets and low voltage this paper could not produce an acceptable copy. *Note* 2—The residual voltage resulting from the second cycle is higher than that from the first charge. From this we may predict that the background of area that is exposed to the second cycle will be darker with reduced contrast.

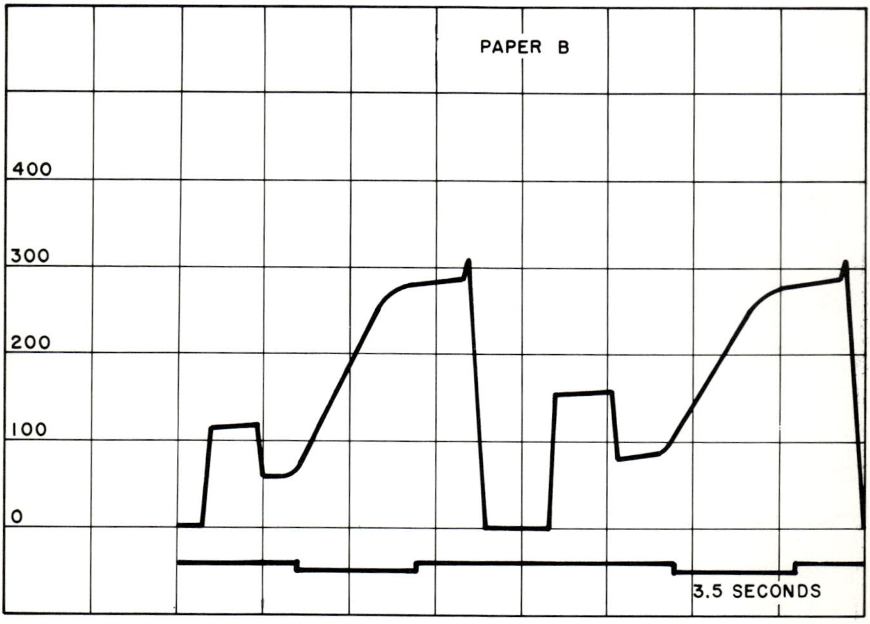

FIG. 7—*Same machine and same 3.5-s exposure as in Fig. 4 but on Paper B.*

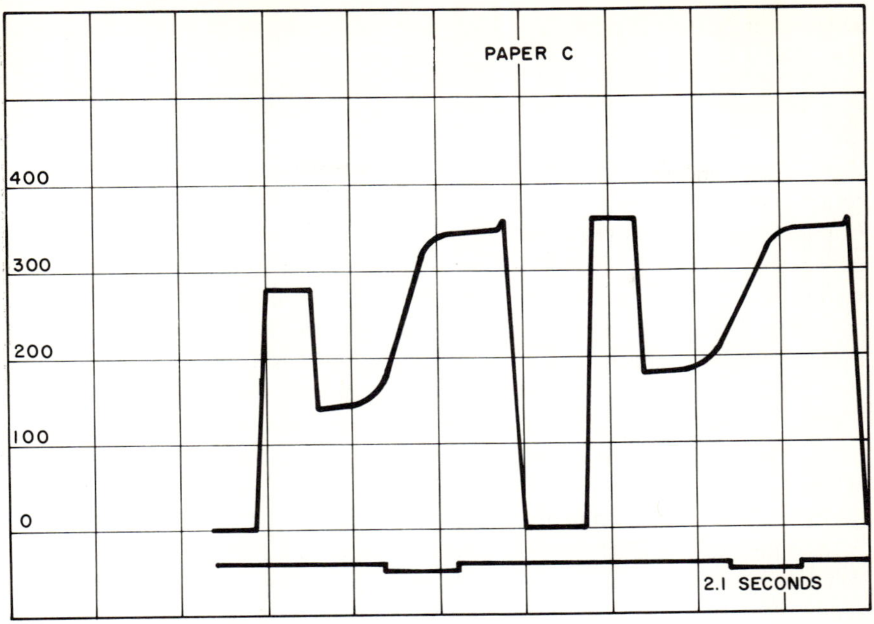

FIG. 8—*Same machine at short exposure on Paper C.*

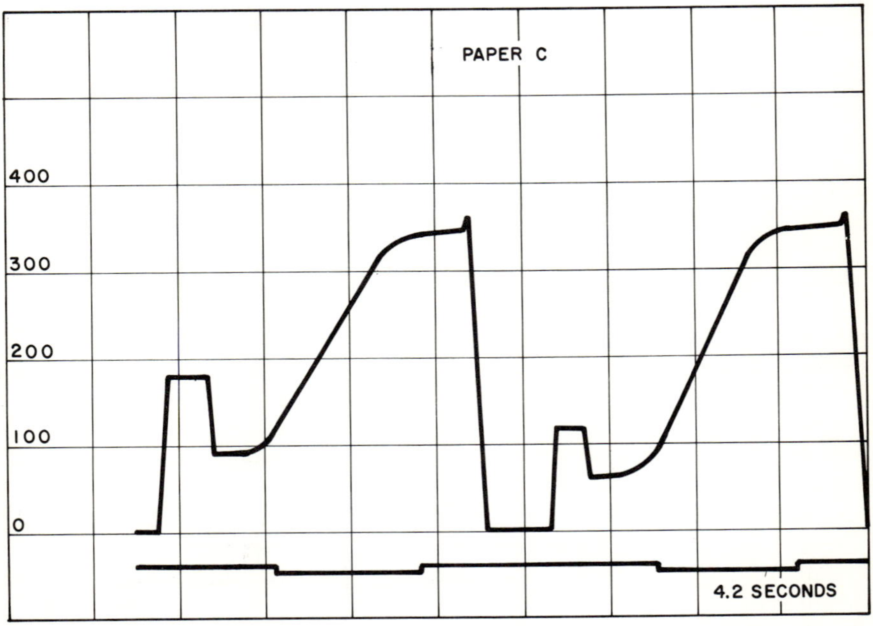

FIG. 9—*Same machine at extended exposure on Paper C.*

Discussion

The instrument evaluation of electrostatic paper from which predictions may be made is only one technique in the total evaluation and may be used only when matched to conditions of the machine on which the paper will be used. For example, Fig. 10 shows Paper A with exposure of 1.4-s. Since residual after first and second cycle are the same, delta is good, and additional exposure is not required (light decay ends as light goes off). We would expect this to be the point of optimum copy and accept the paper. However, we know that this is good paper, but the optimum copy is at 3.5-s exposure from Fig. 4. The only change made was to increase the light to a level higher than normal for the machine. Additionally, at this increased light level a slow paper, not acceptable under normal conditions, would be approved.

Other techniques used in total evaluation are those to determine: (*a*) dry time, (*b*) weight of coating and base stock, (*c*) thickness, (*d*) coating characteristics, (*e*) back marking, (*f*) simulator for copy from roll machines, and (*g*) optical density equipment for copy evaluation.

Instrumental evaluation of electrostatic paper is a technique that will predict optimum copy quality and will explain reasons for some copy characteristics when we recognize that the instrument system must match the machine system and that their predictions hold only for one type of target. Every type of paper and target will change the optimum copy. If we study the machine we will then know what to expect from a target change. While the

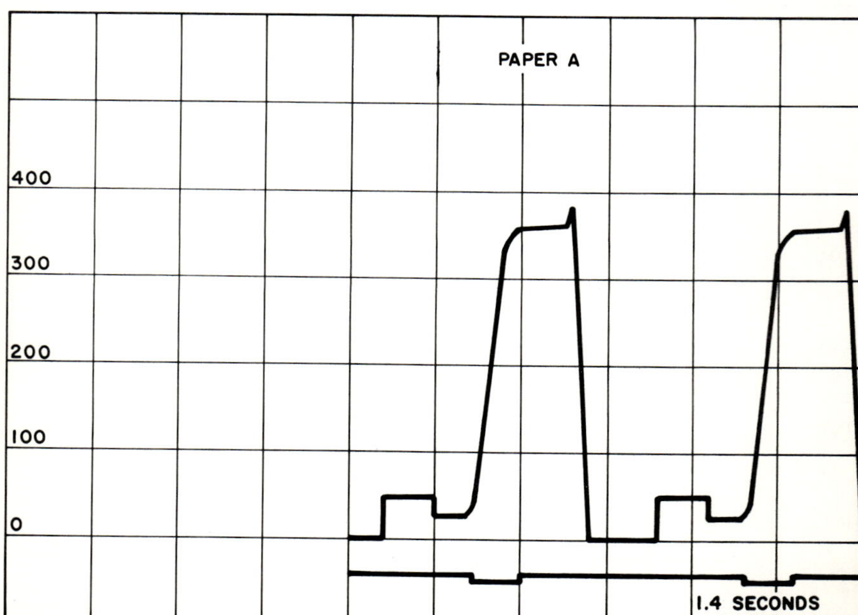

FIG. 10—*Same machine at 1.4-s exposure on Paper A*

copying machine will show the same results, the instruments provide information during production and eliminate expensive roll production.

If this technique is properly used the quality control program will show paper which is in or out of specification limits from small specimens made prior to final process and shipment. It also will aid in the selection of those experimental factors or other unknowns that will be processed for more extensive research and costly production and testing. We have presented a technique that has been tried and has worked. We have tried to avoid listing specific instruments, combinations, or parameters, since the key to the system is to use that combination which meets specific conditions as dictated by the electrostatic paper machine to be evaluated.

M. P. Smoak[1]

Use of Conventional Test Equipment in the Evaluation of Typewriter Supplies

REFERENCE: Smoak, M. P., "**Use of Conventional Test Equipment in the Evaluation of Typewriter Supplies,**" *Evaluation Methods for Business Form Copies, ASTM STP 482*, American Society for Testing and Materials, 1971, pp. 36–43.

ABSTRACT: The evaluation of a typewriter supply product is a complex testing application in which measurements related to many different characteristics of the product are required. The characteristics of interest may be relatively simple, such as the length of a ribbon, or complicated, such as the average useful life of a fabric ribbon. Several common difficulties in defining the measurement objectives and selecting the test equipment are discussed. Examples of applications of commercially available test equipment are given. Some of the instruments discussed include counters, microscopes, force transducers, and reflection photometers. In each case emphasis is placed on the necessity of thoroughly understanding both the measurement requirement and the basic function of the available test equipment so that the equipment may be applied properly. Once the measurement is well defined and the equipment is used properly, the results may be interpreted meaningfully.

KEY WORDS: typewriter ribbons, impact printing, test equipment, evaluation, tests

How would you measure the life of a light bulb? How does a yard stick measure length? These two questions and their rather simple answers illustrate a basic fact of all measurements: measurement is primarily a matter of definition and comparison. (The life of a light bulb obviously is defined as the total time of operation before burning out. A yard stick measures length by comparing an unknown length to a standard length, the yard or fractions of a yard.) The importance of this basic fact often is overlooked in selecting an instrument for a particular measurement task; that is, the measurement requirement may not be defined adequately or the person selecting the instrument may not understand the principle of measurement. The risk of this happening is compounded when many different measurements are required, as in the evaluation of a typewriter supply product. For instance, in testing a fabric ribbon, information is required on its length, width, thickness, cloth construction, degree of inking, and expected useful life. The point to be

[1] Senior associate physicist, Office Products Division, International Business Machine Corp., Lexington, Ky. 40507.

stressed is that a measurement must be well defined in the sense that, in addition to the need for the measurement, both the method of measurement and the proper conditions for measurement must be thoroughly understood and maintained.

Examples relating this idea to supplies evaluation show that once the measurement need has been defined, the problem of selecting a suitable instrument is greatly simplified. There are three distinct classes of measurements to be made. Some characteristics may be evaluated without creating a typed image. Ribbon length or thickness are examples of such nonimage measurements. Other characteristics such as write intensity and voiding, however, require image measurements. If image measurements are made, the device which creates the image must be considered, and measurements on the imaging mechanism may be required. Examples in each area illustrate the necessity of defining exactly what is to be measured.

Nonimage Characteristics

Defining and evaluating the nonimage characteristics of a supply product is relatively simple. Some of the more important include the length, width, and thickness of ribbons, the construction of ribbon fabric, the amount of ink on ribbons and carbon paper, and the tensile properties of plastic base material for carbon ribbons. Figure 1 shows some commercially available laboratory equipment suitable for making such measurements. The tension tester is suitable for obtaining stress-strain curves for plastic films. For determination of ribbon length a measuring wheel one foot in circumference and the revolution counter is useful. Thickness may be measured easily with a dial indicator. The next instrument is a thread counting micrometer used in determining the number of threads per inch in ribbon fabric. The amount of ink deposit on carbon paper and ribbons may be determined with the analytical balance which is readable to one-tenth milligram.

Although such measurements are simple they should be performed with care, considering fully the measurement definition. For example, in measuring the thickness of a compressible material such as a fabric or plastic, it is necessary to specify the amount of load which the measuring instrument will place on the material. (The black cylinder above the dial indicator is the load weight.) In measuring the length of a plastic film ribbon the ribbon should not be stretched. To determine the amount of ink of a film ribbon a known area of inked ribbon is first weighed, then the ink is removed with a solvent, and the plastic base material is reweighed. The amount of ink is determined by subtraction and expressed as ink weight per unit area. If the second weight determination is performed before the material is dry the weight of the base material plus any residual solvent will be recorded.

Further consideration of the simple measuring wheel illustrates the comparative nature of all measurements. To measure length with this device the ribbon is used to drive the measuring wheel without slippage or stretching.

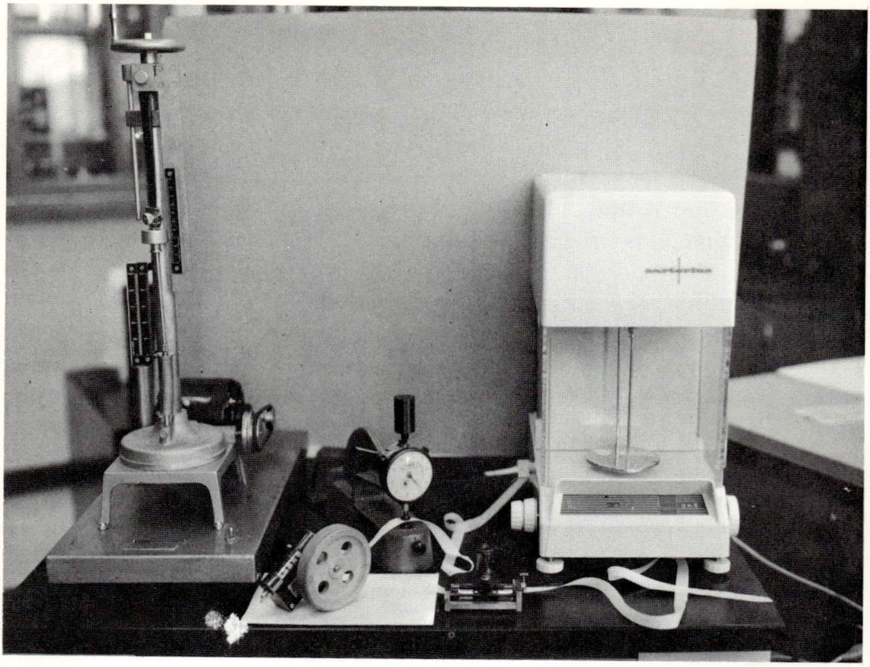

FIG. 1—*Equipment for evaluating supply product nonimage characteristics.* (left to right) *Tension tester, measuring wheel, dial indicator, thread counting micrometer, analytical balance.*

Even if these conditions are met perfectly, however, the resulting measurement can be no more accurate than the accuracy of the measuring wheel circumference. In this case the circumference of the wheel is used as a standard length. In general, measurements obtained with an instrument will be no better than the standards used in calibrating the instrument.

In each of these simple examples the need for a thorough understanding of the measurement and proper measurement conditions is clear. Also, for the measurements to be taken in an absolute sense the instrument must be calibrated with proper standards.

Image Characteristics

Since the printed image is the end result of a typewriter supply product, the characteristics of the image produced are a primary interest in most evaluations. While the need for measurement is recognized the definition of exactly what is to be measured virtually is ignored in many instances. Some of the image characteristics generally considered important include print darkness, uniformity, stroke width, edge sharpness, voiding, extraneous ink, and the life of reusable products. Usually however, additional measurement definitions are required. The term "print darkness" is intuitively meaningful from visual experience but is not defined suitably for measurement; thus, we are led to consider other characteristics such as ink reflectance which may be defined simply as related to print darkness. Again, care in defining what is to be measured greatly simplifies the evaluation process.

For a precise definition of reflectance refer to American Standard Diffuse Reflection Density (Ph 2.17) and ASTM Method of Test for 45-Deg, 0-Deg Directional Reflectance of Opaque Specimens by Filter Photometry (E 97–55). Briefly, reflectance is defined for given geometric and spectral conditions as the ratio of the intensity of light reflected by a specimen area to the intensity of light reflected by a standard magnesium oxide layer. It should be noted that all practical reflection photometers measure the average reflectance over some finite area. Figure 2 shows two reflection photometers. The instrument at left measures the average reflectance of an area 0.008 in. in diameter; the instrument at right measures the average reflectance of an area ½ in. in diameter. Small area reflectometers are useful in measuring ink reflectance within a character stroke. Large area reflectometers may be used to measure the average reflectance of solid or grid patterns such as those shown in Fig. 3. The reflectance reading for a grid pattern is determined by the reflectance of the paper, the reflectance of the ink, and the stroke width of the grid.

Figure 2 also shows a stereomicroscope which is useful in visually examining print specimens and may be equipped with an eyepiece micrometer reticle for making measurements such as stroke width or void size.

An example of well-defined image characteristics may be found in "The Proposed USA Standard Character Set and Print Quality For Optical Character Recognition" (ANSI X3.17). This document defines the image characteristics of importance in optical character recognition (OCR) applications and gives suggested methods of measurement. Print contrast, stroke width, irregularities, spots, and voids are defined. Although the OCR application is highly specialized, many of the definitions in this proposed standard are

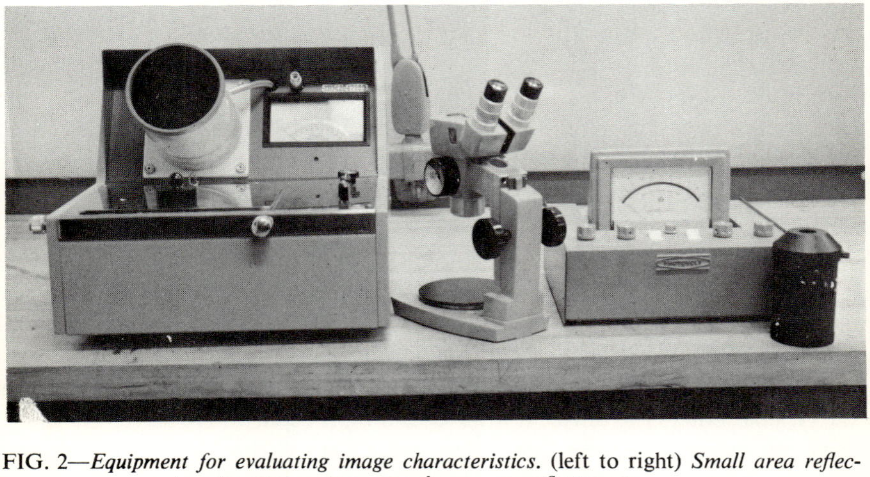

FIG. 2—*Equipment for evaluating image characteristics.* (left to right) *Small area reflectometer, stereomicroscope, large area reflectometer.*

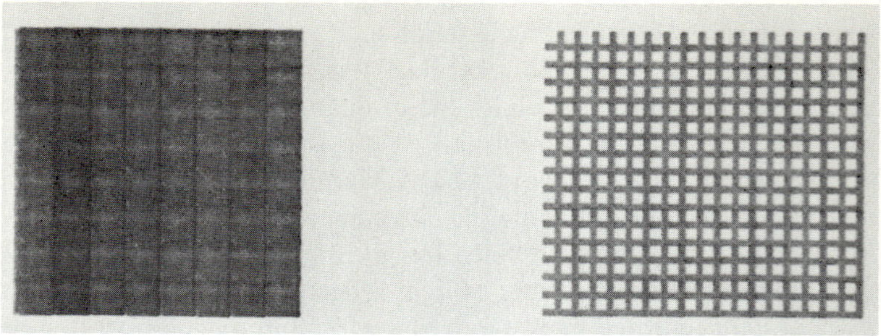

FIG. 3—*Typed specimens for large area reflectometer.*

adaptable to other applications. Measurement of voids, for example, is of general interest in supplies evaluations.

Once the measurements have been defined care must be exercised in maintaining the proper measurement conditions. In particular, instrument modifications, intentional or otherwise, may drastically alter the measurement characteristics of the instrument. For example, modifying the objective lens of the stereomicroscope will change its magnification and hence the calibration of the eyepiece reticle. Reflectometers contain filters to establish the proper spectral conditions. Improper filter substitution has led to confusion on a number of occasions. Again, it should be remembered that the standards to which an instrument is calibrated are of utmost importance. Recalibration of standards or acquisition of new standards should be scheduled at frequent intervals.

Imaging Mechanism Measurements

The print mechanism used to create the test images must be selected and maintained with care. Several criteria should be considered. First, the mechanism should simulate as closely as possible the actual use conditions to which the supply product will be subjected. Second, the mechanism should be consistent and reliable in its impact characteristics. In practice a variety of devices are used including standard office typewriters and devices based on falling weights and pendulums. After selecting an imaging device the primary consideration is guaranteeing day-to-day and test-to-test consistency. Several measurements useful in checking consistency can be defined.

One such measurement is the momentum or energy of the impacting member just before impact—on a typewriter it is sufficient to measure the angular velocity of the typebar just before it contacts the platen. Figure 4 shows a suitable arrangement using an electronic counter as a time interval meter. The counter is controlled by the device mounted at the type guide. As the typebar enters the device it interrupts a light beam which starts the timer, then interrupts a second light beam which stops the timer. The typebar velocity may be calculated from the time measurement and the distance

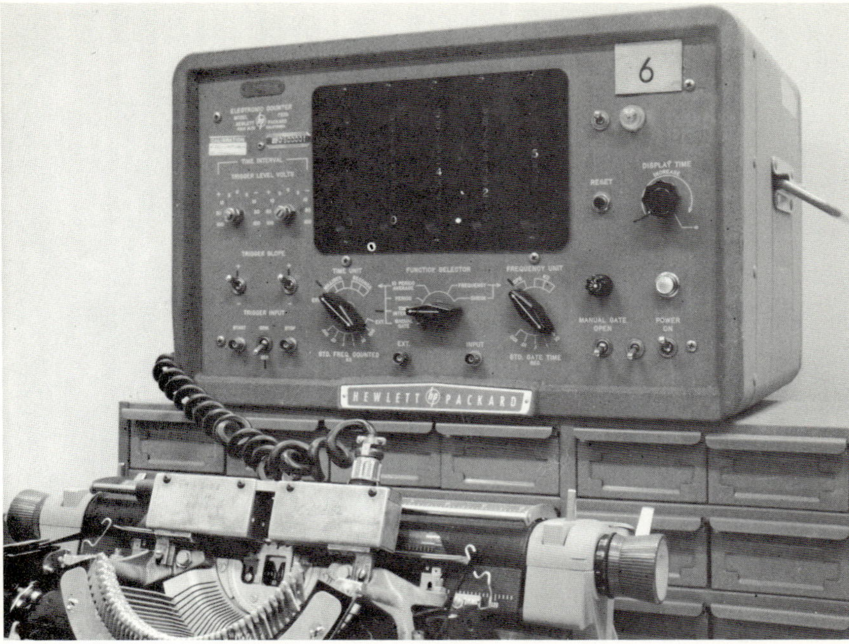

FIG. 4—*Electronic time interval meter and photoelectric control unit for measuring typebar velocity.*

between the two light beams. This measurement can be adapted easily to other printing mechanisms.

Another approach is to measure the force produced by the impacting member using a force transducer, a device which converts force to an electrical voltage which may be displayed on an oscilloscope. Figure 5 shows a device suitable for making force measurements on a typewriter. It consists of a special platen for mounting the device, a crystal force transducer with a rubber impacting surface, an amplifier, and an oscilloscope. In use, the force transducer is struck by the typebar and produces a voltage signal on the oscilloscope corresponding to actual impact forces. Figure 6 shows a force signal recorded on the oscilloscope. The ordinate can be calibrated in force units. The abscissa is in time units. It should be noted that the force signal produced depends greatly on the type and hardness of the rubber used on the impacting surface of the force transducer; thus, care should be exercised in selecting and maintaining the rubber surface. Variations of this technique may be applied to other printing mechanisms.

A third technique used for checking print mechanism consistency is the measurement of embossing. On some print mechanisms the forces created are great enough to emboss aluminum foil to a degree measurable with a dial indicator or micrometer. If this technique does not seem applicable a

FIG. 5—*Force measuring equipment.* (left to right) *Force transducer mounted in modified platen, amplifier, oscilloscope.*

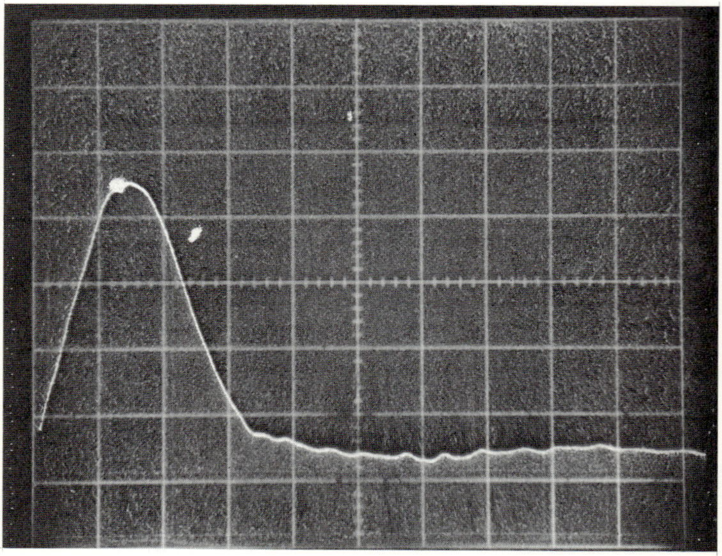

FIG. 6—*Oscilloscope trace of force during typebar impact.*

light section microscope or scratch measuring microscope may be used to measure the actual embossing on smooth paper.

These measurements are suggested for checking the consistency of a particular mechanism and may not be adequate for comparing different mechansims, as when two typebars with identical angular velocity may print differently because of differences in their alignment with the platen. As with other measurements, proper conditions and calibrations must be maintained if the measurements are to be interpreted meaningfully.

Summary

The examples given have shown that measurements must be defined thoroughly, and the proper conditions for measurement must be understood and maintained. In addition the comparative nature of all measurements is stressed, pointing out the importance of calibration standards. These points should be kept foremost in mind when selecting test equipment or establishing test methods. Failure to understand the method of measurement can result in the measurement being misinterpreted. Failure to maintain the proper measurement conditions can result in data which is meaningless.

G. T. Slaughter[1]

Mechanics of Ink Transfer During Impact Printing

REFERENCE: Slaughter, G. T., **"Mechanics of Ink Transfer During Impact Printing,"** *Evaluation Methods for Business Form Copies, ASTM STP 482*, American Society for Testing and Materials, 1971, pp. 44–55.

ABSTRACT: Discerning customers and more rigorous applications of impact imaged documents demand better understanding of ink release from typewriter ribbons. Considerable effort has been expended toward this end; unfortunately, the results of this effort have not been utilized fully by printing mechanism designers and ink formulators. For this reason an attempt is made to bridge the gap between mechanical and chemical variables by addressing those parameters affected by both.

Regarding ink transfer from carbon or film ribbons, the transfer mechanism involves brittle fracture at character edges and release by generation of a shear stress at the ink substrate interface or within the ink film, depending upon the physical nature of the film. Evidence supporting this theory has been collected through literature search, observation during printer and ribbon development, and from measurements of ink release stresses.

KEY WORDS: impact printing, ink transfer, typewriter ribbons, shear stress, paper elasticity, impact dynamics, evaluation, tests

Several explanations of carbon ink release from plastic substrates have been proposed. These explanations have been proposed in such a way as to render them of little value in actual mechanical design and ribbon formulation. Notable examples are the brittle fracture and thermal fusion theories.

The brittle fracture theory[2] proposes that, under compression, a carbon ink film is shattered. The resultant particles adhere to the paper through the action of an oil or other vehicle which exudes from the ink film under influence of the impact pressure.

The thermal fusion theory[3] proposes that sufficient heat is generated internal to the ink film, during impact to cause softening of the ink. Ink, thus made more fluid, is released from the film and adheres to the paper.

[1] Senior engineer, Office Products Division, International Business Machine Corp., Lexington, Ky. 40507.

[2] Gates, R. G. and Landzberg, A. H., "The Transfer of Solid Ink in Impact Printing, Part I," Research Report RC-340, IBM Corp., New York, Nov. 1960.

[3] Hso, C. S. and Anderson, A. G., "Investigation of Typewriter Printing," IBM 05.041.374, IBM Corp., New York, 12 May 1952.

The shear theory proposes that the ink film releases from the plastic substrate under the influence of a shear stress generated at the interface. The mechanics of generation of this shear stress will be examined in this paper. Once released, the ink film adheres to the paper through the influence of vehicles or "wetting agents" or simple mechanical interlocking to the paper topography or both.

The variables contributing to shear are not quantified. The intent is to examine these variables in principle with the result that impact printing mechanism and ribbon designers shall know which variables are important to their particular case. Variables are examined in sufficient detail so as to identify cause and effect relationships in ink transfer.

Discussion

Shear Release

Figure 1 compares two attempts to print the same character under identical conditions except that in Fig. 1*B* a gross shear was introduced by weakening certain members in the print mechanism so as to cause a force in the horizontal direction as a result of the normal printing force.

The impulse curves for each case are displayed beneath the corresponding characters. It should be noted that the normal force was damped substantially by this mechanical weakening. Obviously, ink release is improved by generation of a shear force even though the peak normal force is reduced as shown.

Evidence supporting the theory of shear release was gathered by static measurement. Three square pads with a total face area of 1.81×10^{-2} cm^2 were machined on a steel block. The block was applied to a printing laminate comprising an aluminum plate, single sheet of paper, and sheet of carbon film typewriter ribbon. Weights were applied statically to create a normal stress of 54.5×10^5 kg/m^2. No ink was transferred. The block was then attached to a tension tester, and a horizontal load was applied at the rate of 1.3 cm/min. It was found that ink first transferred completely at a normal stress of from 8.1×10^5 to 33.4×10^5 kg/m^2 when accompanied by a shear stress of 1.7×10^5 to 3.5×10^5 kg/m^2, varying between these limits as a function of the paper used.

As the normal stress was increased, the shear stress necessary for transfer increased. Later testing showed that the same characteristic existed when no ink was used. That is, the ribbon was replaced by inkless substrate. It was concluded that the normal stress caused compression of the paper, and the shear force was opposed by the paper exerting force against the sides of the characters as well as being opposed by the friction resistance of the ribbon. This being the case, the shear stress values at release do not truly represent the stress as seen by the ribbon. However, they do represent the values which

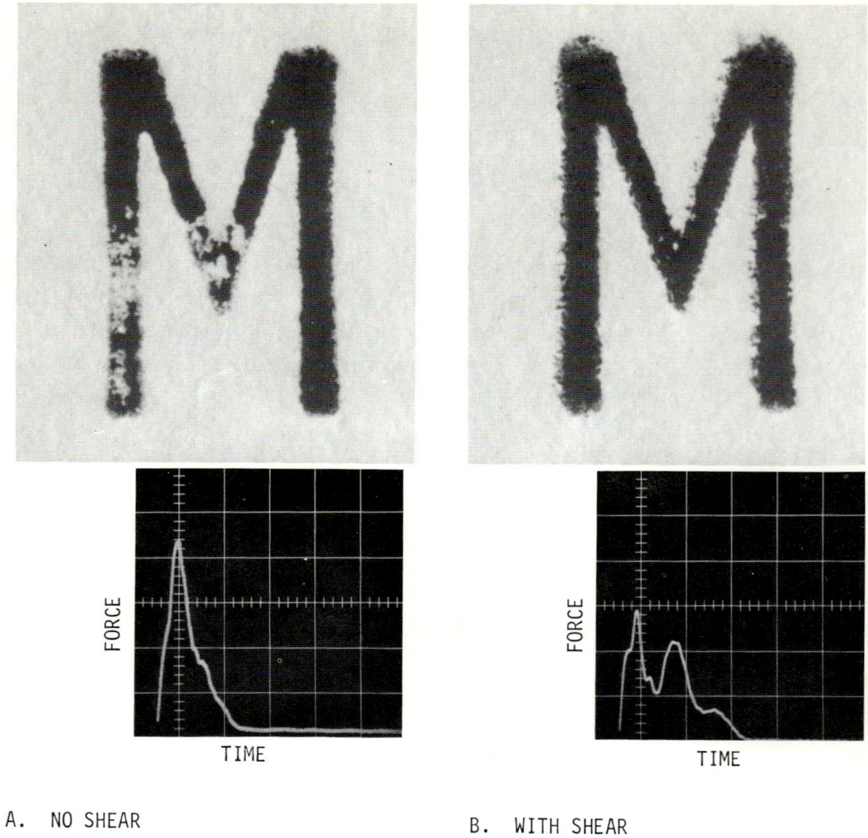

FIG. 1—*Effect of shear force on ink release.*

must be generated by the impact. Figure 2 shows the shear stress for ink release as a function of the normal stress. Paper A was a 28-lb ledger. Paper B was a plastic film, surface treated so as to be ink receptive.

Given evidence that shear is at least an important contributor to the mechanism of ink release, it is appropriate to examine the contribution made by each component of the printing system. In order to do this, reference to Fig. 3 will be of some assistance. The figure is a schematic depiction of the printing laminate, and each component is identified.

Mechanics for Shear

Consideration is first given to the elementary principles of mechanics which must persist if the shear release mechanism is active. The assumptions are made that the paper is fixed to the platen and that some means of generating a shear component, S_S, is accomplished. Also assume that some critical shear stress, S_C, exists at which the ink to substrate bond is broken.

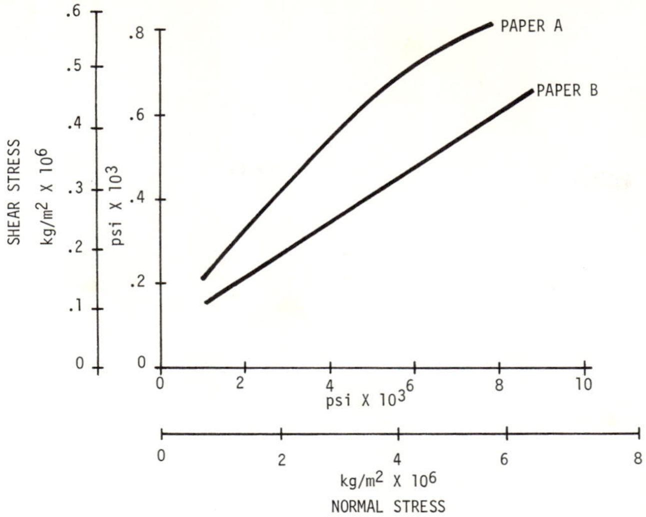

FIG. 2—*Shear stress at release versus normal stress.*

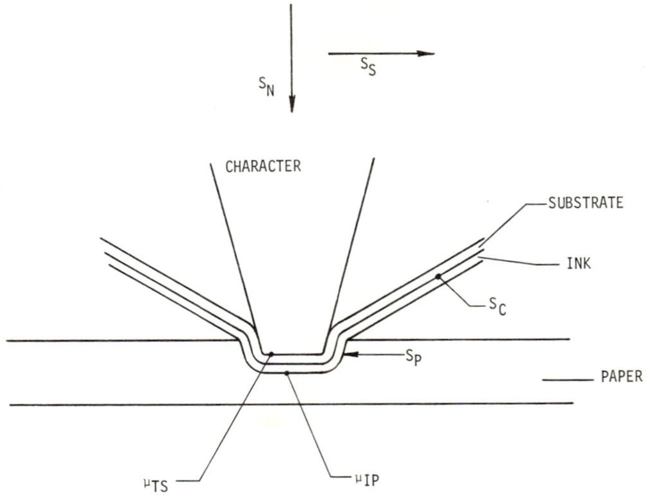

FIG. 3—*Schematic of printing laminate.*

It is first necessary that the products of the normal stress, S_N, and both friction coefficients, μ_{TS} and μ_{IP}, be greater than S_C. If this were not the case any shear force over and above that necessary to overcome paper resistance would cause slipping at the interface with the lower friction coefficient. If μ_{IP} is lower, ink smear obviously will result to the extent that the character face moves. If μ_{TS} is lower, smear will not occur, but neither will ink release. A second requirement for transfer is that S_S be equal to or greater than the sum of S_C and the paper resistance, S_P. This in turn defines an interaction between S_S and S_N because S_P is a function of S_N.

In order to define this function, refer to Fig. 4. A coated, photomaster paper was used to generate both lines. The top line was generated as described for Fig. 2. The bottom line was generated in the same way except that no ink film was present. An inkless substrate was used instead. The lower curve represents the paper resistance, S_P, as a function of the normal force, S_N, for the particular character configuration used in this test. The difference between the two lines is the ink release stress, S_C, for the particular ribbon used. This difference should be constant, but experimental error causes some variation.

It is important to note that the result is characteristically as expected. S_P is dependent on embossing so Fig. 4 could be viewed as a compression stress strain relationship for paper. As the paper fibers are compacted one would expect the rate of strain to decrease.

The limitations of Fig. 4 should be noted. The influence of such dynamic factors as rate of stress is ignored. This influence is significant if the mechanism of ink to paper adhesion is through exudation of oil or vehicle toward the paper surface under compression of the ink. The character configuration used imposes a limit because the relationship between S_P and S_N cannot be extrapolated directly to characters of different configurations. This will be discussed later. Also, the release stress of the ribbon used would not remain unchanged if substantially different materials were used to construct the ribbon.

Ribbon Properties

The influence on print quality contributed by the ribbon can now be discussed in principle. In the first case, assume that the ink film is substantially continuous, has relatively high cohesive strength, and is brittle relative to the substrate. Under influence of an impact wherein forces such as described in Fig. 3 exist, the following events will occur.

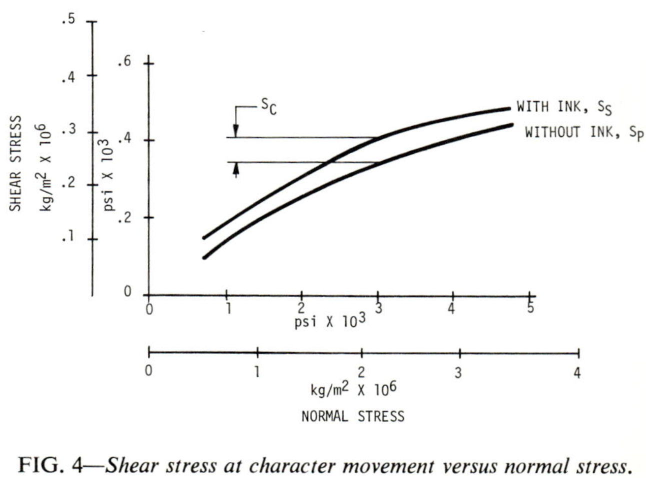

FIG. 4—*Shear stress at character movement versus normal stress.*

As the normal stress increases, embossing will occur. Embossing will cause high bending stresses in the ink film at the character edges. These stresses will cause the ink film to fracture. Edge definition of the printed character is a function of irregularity of the line of fracture. This is dependent on the following salient parameters: angle of the character face with the normal to the paper surface, topography of the paper surface, and existence of macroscopic particles in the ink which are not easily broken.

The shear stress will cause a shearing action between the substrate and ink film. As discussed previously, this stress must be above some critical value, S_C, at all points within the character face to effect complete transfer. S_C is controlled by the choice of materials in the ribbon. If, for example, a component of the ink acts as a plasticizer for the substrate the release stress could be made extremely high. When the shear stress reaches a value equal to the sum of paper resistance, corresponding to normal stress at that time of the impact (refer to Fig. 4), and the ribbon release stress, the ink will break from the substrate. Mechanical adhesion causes the ink to stick to the paper. The resulting image is characteristic, as shown in Fig. 5. Voids, if present, are generally visible to the eye, and character edges may contain large ink particles loosely attached to the paper. The used ribbon also is displayed in Fig. 5.

In the second case, assume that the nature of the ink film is particulate. That is, the ink comprises relatively discreet particles with low cohesion. Under impact the following mechanism is proposed.

As embossing progresses, small, local relative motion between ink and paper will exist at the character edges. Particles will be sheared from the ink film. Immediately at the character corner, where relatively high normal stresses exist, the density of particles transferred will be high. The density of

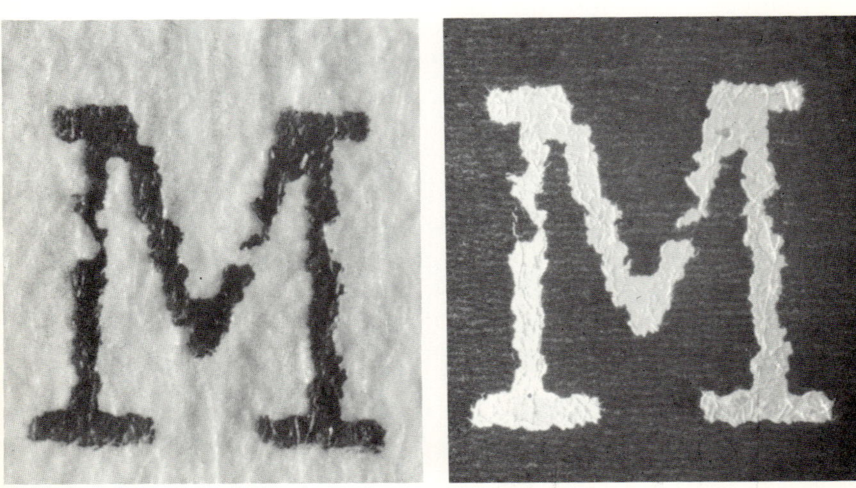

FIG. 5—*Character and used ribbon having strong ink film.*

transferred particles will decrease toward the point of last contact between ribbon and paper. Character edges thus will have the appearance of a gradual transition from black to gray.

Shear stress will cause dislocation of particles inside the character face after paper resistance is overcome. The active controlling ribbon parameter is the cohesive force existing between ink particles. In this case there is no well-defined plane of fracture. Character voids are typically microscopic and may be distributed fairly uniformly such that the entire character appears to be gray, as shown in Fig. 6.

Substrate plasticity is important to print quality. It is possible to envision a substrate substantially rigid so as to deter brittle fracture at character edges described in Case 1 above. This same rigidity also could preclude intimate contact between ink and paper in localized areas. This will be examined further in the discussion of paper contributions to print quality.

In summary, the ribbon parameters found to be important to ink release are cohesive strength, ink to substrate bond, ink brittleness, and substrate plasticity. Those parameters alluded to be important to adhesion to paper comprise vehicle content and amorphousticity of the ink as well as substrate plasticity.

Character Geometry

Those factors in character configuration influencing print quality include the angle at the character edge, face area, total line segment length perpendicular to the direction of shear, line width of the type face, and surface condition of the character face.

The angle at the character edge and the radius of the character corner help determine edge definition, as discussed previously. There are, obviously,

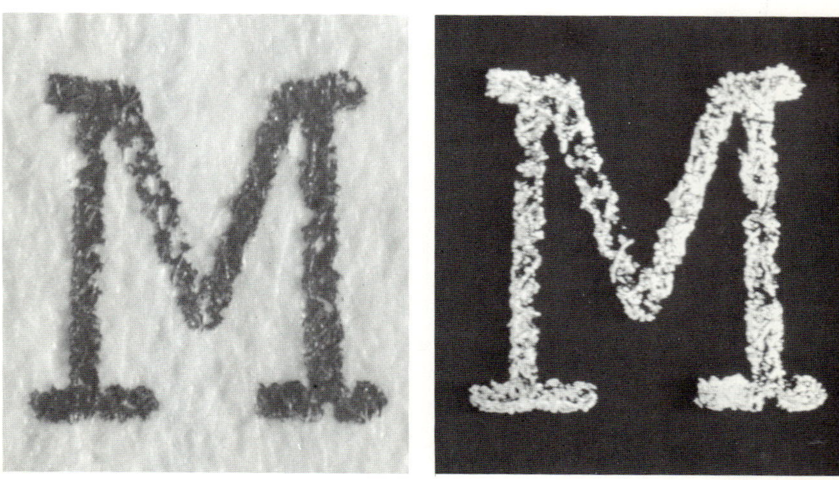

FIG. 6—*Character and used ribbon having particulate ink.*

practical limits imposed on these parameters by the manufacturing methods used to construct the type, but both should be minimized to optimize edge definition. There is a functional disadvantage to having the character edge extremely sharp. Cutting of the ribbon and paper are aggravated. The reason for this cutting can be understood by the following mathematical model.

M. Sadowsky determined the distribution of pressure under a rigid plate of width, $2b$, under a plane strain for an elastic material. Assuming a plane of infinite length, having the edges at $x = \pm b$, he found that the normal stress under the plate is distributed according to the function.[4]

$$\sigma_Y = - \frac{Q}{\pi\sqrt{b^2 - x^2}} \text{ for } - b < x < b \qquad (1)$$

where Q denotes the normal force in kilograms per meter exerted by the plate in the area $2b$ by one meter.

Equation 1 can be written as follows.

$$\sigma_Y \frac{b}{Q} = - \frac{1}{\pi\sqrt{1 - (x/b)^2}} \qquad (2)$$

Equation 2 is plotted in Fig. 7.

This mathematical model, though idealized, clearly explains why the edges of a character emboss more than the center. This principle applies to a section of a long character bar such as the slash or diagonal.

The slash was impacted, without a ribbon, against a direct typing offset master. A *Zeiss* light section microscope was used to measure permanent embossing of the master. The average embossing was 0.045 mm at the ends of the slash and 0.023 mm at the vertical center. The face radius matched the radius of the paper, that is, there was mechanical fit between the character and paper.

It is evident that, in actual practice, the normal stress distribution is somewhat altered from Fig. 4. This alteration is a result of the character edge condition, as discussed previously. The principle does apply, and it is enlightening to note that over 75 percent of the character line width the normal stress is less than 50 percent of the average stress as calculated by force divided by area. This principle is important to ink release.

Consider the crossbar in a boldface character H. Assuming that the mathematical model is correct and ignoring end conditions, the average normal stress along a vertical center line through the center of the crossbar is 0.318 times total force over total area. In addition, a section through the crossbar at this center line would again reduce the stress at the geometric center of

[4] Nadai, A., *Theory of Flow and Fractures of Solids–Volume II*, McGraw-Hill, New York, 1963, p. 224.

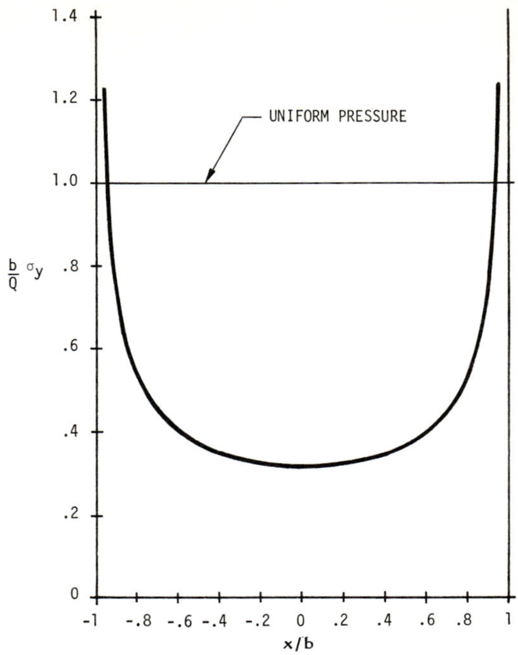

FIG. 7—*Normal stress distribution under a rigid plate.*

the crossbar by a factor of 0.318. These brute force assumptions lead to the conclusion that the local normal stress is an order of magnitude lower than the average normal stress. The product of the normal stress and friction coefficients easily could be lower than the release stress of the ribbon at a point near the character's geometric center even though the force over area computation shows that it is not.

Obviously, face area of a character is important in determining ink transfer. However, it has been shown that the configuration or shape of the character is also important. In the discussion of paper resistance to shear it was noted that the depth of penetration was a controlling variable. It follows that the total projected length of character lines in a direction perpendicular to the direction of shear is an important variable. Although no empirical data has been taken, one would expect that paper resistance would approximate a linear function of this length. Figure 8 explains the concept of total projected length. The length in question is the sum of L_1 and L_2 if the shear direction is upward as shown.

It is theoretically possible to generate a shear stress, adequate to effect transfer, without the presence of a lateral force. If the character face dynamically deforms the ribbon and paper so as to cause elongation of the ribbon substrate across the total character line width, and, if the ink is substantially less elastic than the film, the ink will be released by the stretching action of

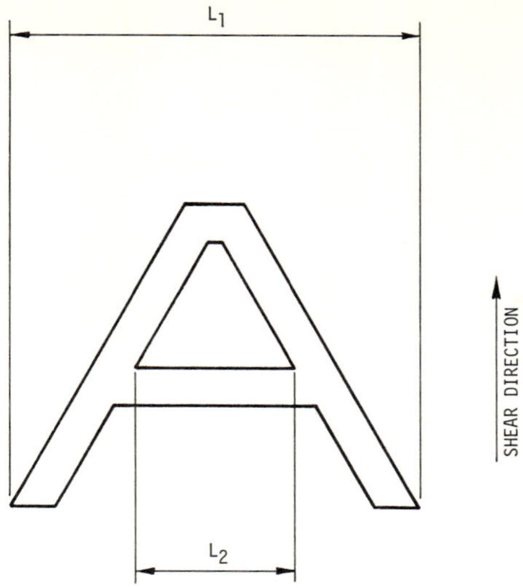

FIG. 8—*Projected length opposing shear motion.*

the film. Character conditions which enhance this mechanism are narrow line segments (fineline type) and well-rounded character corners.

The surface condition of the typeface is important to ink release because it controls the friction coefficient between the typeface and ribbon substrate. The limit on typeface roughness is the point at which the surface texture begins to appear to the naked eye.

Paper Properties

Paper contributes to the mechanism of ink release through the influence of its topography and elasticity. It influences print quality further by its surface area and wettability.

The conditions of shear for ink release require normal stress levels equal to the ratio of the ribbon release stress to the lower friction coefficient in the printing laminate of Fig. 4. If the paper surface is rough and rigid, the conditions of Fig. 6 will persist. That is, numerous voids, emphasizing the paper topography, will appear in the character. These occur as a result of inadequate normal stress in low spots.

The paper topography also affects character edge definition. The ink film preferentially breaks at high spots in the paper.

Paper elasticity has an additional effect on ink release. A highly elastic paper (or manifold pack) reduces the normal force attained during impact. This reduction can cause inadequate normal stress for ink release. Figure 9 compares normal force versus time of a character with one sheet of plastic drafting film with that of the same character with a 20-lb bond original and a five-copy manifold pack. This result easily is explained by visualizing an

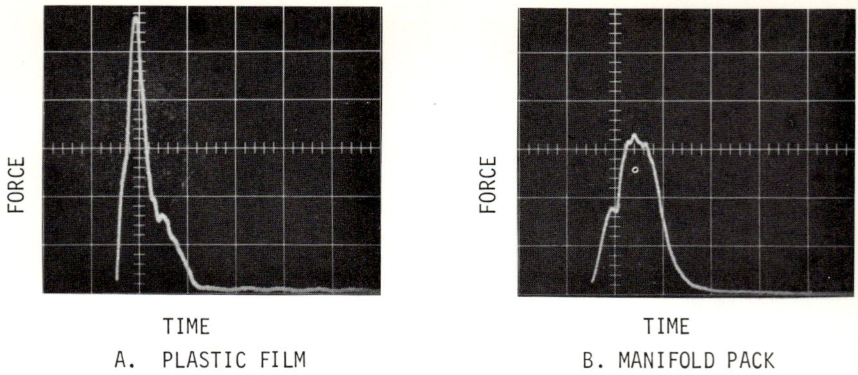

FIG. 9—*Effect of paper on maximum print force.*

additional low rate spring imposed in series with an otherwise high rate system.

An impervious paper will not accept ink as well as a porous or granular surface. A most obvious physical embodiment characterizing wettability and total surface area is found in Fig. 10. The paper of Fig. 10*A* is an erasable bond. It is made erasable by surface treatment with a material which acts as a barrier. The paper is characteristically nonabsorbent to oils and has low total surface area. Figure 10*B* is a highly receptive, clay coated paper.

Print Mechanics

The printing mechanism controls the generation of a shear stress. There are at least two methods of shear stress generation which merit consideration.

First, if the momentum vector of the printing mass at impact is at an angle to the normal to the paper, a shear force will be generated. Obviously, planar

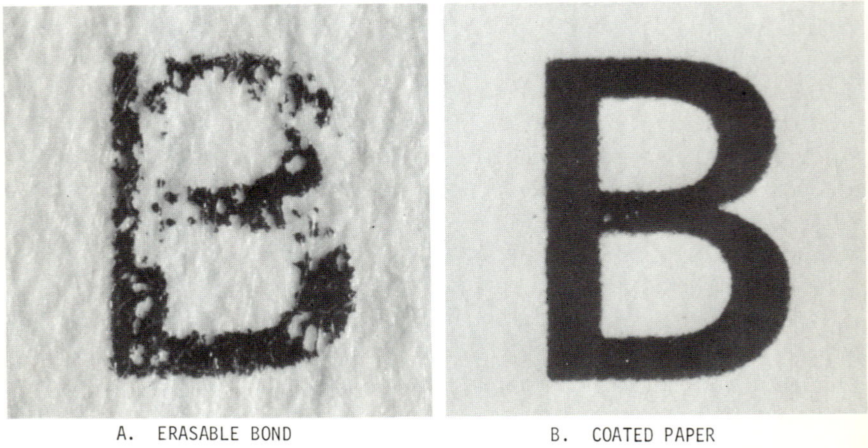

FIG. 10—*Effect of paper surface on character coverage.*

match between the typeface and paper must be maintained. The shear and normal forces will be related by the appropriate trigonometric functions through the print impulse. Second, a shear force can be generated by elastic bending of a type carrier bar, for example, in a plane normal to the paper plane. In this case, the shear and normal forces are related by the geometry and elasticity of the bar.

It will be obvious to one skilled in the art that the conditions for transfer must exist simultaneously, that is, at the same time during the print duration. The shear and normal stresses must bear the relationship described by Figs. 2 and 4 at some time during the impulse. The relationship is a function of ribbon, paper, and character.

General Summary

The principles involved in the effect of each component of a printing system on ink release have been discussed. A simplified summary of these principles is given below in equation form using the nomenclature of Fig. 4.

Requirements for ink transfer

S_N must cause complete ink to paper contact
$\mu_{TS}\, S_N$ and $\mu_{IP}\, S_N > S_C$
$S_S > S_P + S_C$

Source of variable data

μ_{IP} = constant—measured
μ_{TS} = constant—measured
S_P = $f(S_N)$—empirical data as in Figs. 2 and 4
S_C = constant—empirical data as in Figs. 2 and 4

This allows description of a functional relationship between S_S and S_N useful to the printer design.

Acknowledgments

The author extends appreciation to R. D. Mathews who has contributed greatly to the science of impact printing. The efforts of Department 524 personnel who contributed to the data presented are gratefully acknowledged.